EXPERIENCING
GEOMETRY

GEOMETRY

Logic can only go so far –
    after that I must
see-perceive-imagine.
This geometry can help.

I may reason logically thru theorem
    and propositions galore,
      but only what I perceive is real.

If after studying I am not changed –
    if after studying I still see the same –
      then all has gone for naught.

Geometry is to open up my mind
  so I may see what has always been
behind
      the illusions that time
        and space construct.

Space isn't made of point and line
    the points and lines are in the mind.
The physicists see space as curved
    with particles that are quite blurred.
And, when I draw, everything is fat
    there are no points and that is that.
The artists and the dreamer knows
    that space is where an image grows.
For me it's a sea in which I swim
    a formless sea of hope and whim.

Thru my fear of Infinity and One
    I structure space to confine
      my imagination away from the idea
        that all is One.

But, I can from this trap escape –
    I can see the geometry in which I
wander
    as but a structure I made to ponder.

I can dare to let go the structures
            and my fears
    and look beyond
  to see what is always there to see.

But, to let go, I must first grab on.
    Geometry is both the grabbing on
      and the letting go.
It is a logical structure
      and a perceived meaning –
      Q.E.D.'s and "Oh! I see!"'s.
It is formal abstractions
    and beautiful contraptions.
It is talking precisely about that
    which we know only fuzzily.
But, in the end, and, most of all,
    it is seeing-perceiving
      the meaning that
        I AM.

— David Henderson, 1978

EXPERIENCING GEOMETRY

IN EUCLIDEAN, SPHERICAL, AND HYPERBOLIC SPACES

SECOND EDITION

David W. Henderson
Cornell University, Ithaca, New York

Contributor
Daina Taimiņa
University of Latvia, Rīga

Contributors to the First Edition
(*Experiencing Geometry on Plane and Sphere*)
Eduarda Moura
Justin Collins, Kelly Gaddis, Elizabeth B. Porter, Hal W. Schnee, Avery Solomon

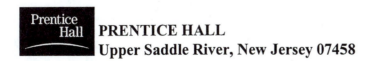
PRENTICE HALL
Upper Saddle River, New Jersey 07458

Library of Congress Cataloging-in-Publication Data

Henderson, David W. (David Wilson), 1939–
    Experiencing geometry: in Euclidean, spherical, and hyperbolic spaces.--2nd ed./
David W. Henderson.
        p. cm.
    Includes bibliographical references and index.
    ISBN 0-13-030953-2
    1. Geometry. I. Title: Experiencing geometry in Euclidean, spherical, and hyperbolic
spaces. II. Title.

QA453 .H497 2001
516--dc21   00-037494

Acquisitions Editor: George Lobell
Assistant Vice President of Production and Manufacturing: David W. Riccardi
Executive Managing Editor: Kathleen Schiaparelli
Senior Managing Editor: Linda Mihatov Behrens
Production Editor: Betsy Williams
Manufacturing Buyer: Alan Fischer
Manufacturing Manager: Trudy Pisciotti
Marketing Manager: Angela Battle
Marketing Assistant: Vince Jansen
Director of Marketing: John Tweeddale
Editorial Assistant/Supplements Editor: Gale Epps
Art Director: Jayne Conte
Cover Designer: Joseph Sengotta
Cover Image: Paul Chesley/Stone; cover photo is of the Pantheon in Rome

© 2001, 1996 by Prentice-Hall, Inc.
Upper Saddle River, NJ 07458

Portions of this material are based upon work supported by the National Science Foundation under
Grant No. USE-9155873 and by Dwight David Eisenhower Title IIA grants administered by the
New York State Department of Education. Any opinions, findings, and conclusions or recommen-
dations expressed in this material are those of the author and do not necessarily reflect the views of
the National Science Foundation or the New York State Department of Education.

Printed in the United States of America
10 9 8 7 6 5 4 3 2 1

ISBN 0-13-030953-2

Prentice-Hall International (UK) Limited, *London*
Prentice-Hall of Australia Pty. Limited, *Sydney*
Prentice-Hall of Canada, Inc., *Toronto*
Prentice-Hall Hispanoamericana, S.A., *Mexico*
Prentice-Hall of India Private Limited, *New Delhi*
Prentice-Hall of Japan, Inc., *Tokyo*
Pearson Education Asia Pte. Ltd.
Editora Prentice-Hall do Brasil, Ltda., *Rio de Janeiro*

To all the students who have studied geometry with me
(you have taught me much about geometry)

and to

Mount Ætna, the śḳūnītis, and the angels

CONTENTS

PREFACE

What geometrician or arithmetician could fail to take pleasure in the symmetries, correspondences and principles of order observed in visible things? Consider, even, the case of pictures: those seeing by the bodily sense the productions of the art of painting do not see the one thing in the one only way; they are deeply stirred by recognizing in the objects depicted to the eyes the presentation of what lies in the idea, and so are called to recollection of the truth — the very experience out of which Love rises.
— Plotinus, *The Enneads*, II.9.16 [**A:** Plotinus]

This book is an expansion and revision of the book *Experiencing Geometry on Plane and Sphere*. The most important change is that I have included material on hyperbolic geometry that was missing in the first book. This has also necessitated more discussions of circles and their properties. In addition, there is added material on geometric manifolds and the shape of space. I decided to include hyperbolic geometry for two reasons: 1) the cosmologists say that our physical universe very likely has (at least in part) hyperbolic geometry, and 2) Daina Taimiṇa, a mathematician at the University of Latvia and now my wife, figured out how to crochet a hyperbolic plane, which allowed us to explore intuitively for the first time the geometry of the hyperbolic plane. In addition, Daina Taimiṇa has been responsible for including in this edition significantly more historical material. In this historical material we discuss and try to clear up many current misconceptions that are commonly held about some mathematical ideas.

This book is based on a junior/senior-level course I have been teaching since 1974 at Cornell for mathematics majors, high school teachers, future high school teachers, and others. Most of the chapters start intuitively so that they are accessible to a general reader with no particular mathematics background except imagination and a willingness to struggle with ideas. However, the discussions in the book were written for mathematics majors and mathematics teachers and thus assume of the reader a corresponding level of interest and mathematical sophistication. Certain problems and sections in this book require from the reader a background more advanced than first-semester calculus. These sections are indicated with an asterisk (*) and the background required is indicated (usually at the beginning of the chapter).

The course emphasizes learning geometry using reason, intuitive understanding, and insightful personal experiences of meanings in geometry. To accomplish this the students are given a series of inviting and challenging problems, and are encouraged to write and speak their reasonings and understandings. I listen to and critique their thinking and use it to stimulate whole class discussions.

The formal expression of "straightness" is a very difficult formal area of mathematics. However, the concept of "straight" an often used part of ordinary language, is generally used and experienced by humans starting at a very early age. This book will lead the reader on an exploration of the notion of straightness and the closely related notion of parallel on the (Euclidean) plane, on a sphere, or on a hyperbolic plane. We will study these ideas and questions, as much as is possible, from an intrinsic point-of-view — that is, the point-of-view of a 2-dimensional bug crawling around on the surface. This will lead to the question: "What is the shape of our physical three-dimensional universe?" Here we are like 3-dimensional bugs who can only view the universe intrinsically.

Most of the problems are approached both in the context of the plane and in the context of a sphere or hyperbolic plane (and sometimes a geometric manifold). I find that by exploring the geometry of a sphere and a hyperbolic plane my students gain a deeper understanding of the geometry of the (Euclidean) plane. For example, the question of whether or not Side-Angle-Side holds on a sphere leads one to pursue the question of what is it about Side-Angle-Side that makes it true on the plane. I also introduce the modern notion of "parallel transport along a geodesic," which is a notion of parallelism that makes sense on the plane but also on a sphere or hyperbolic plane (in fact, on any surface). While exploring parallel transport on a sphere the students are able to more fully appreciate that the similarities and differences between the Euclidean geometry of the plane and the non-Euclidean geometries of a sphere or hyperbolic plane are not adequately described by the usual Parallel Postulate. I find that the early interplay between the plane and spheres and hyperbolic planes enriches all the later topics whether on the plane or on spheres and hyperbolic planes. (All of these benefits will also exist by only studying the plane and spheres for those instructors that choose to do so.)

USEFUL SUPPLEMENTS

A faculty member may obtain from the publisher the *Instructor's Manual* (containing possible solutions to each problem and discussions on how to use this book in a course) by sending a request via e-mail to George_Lobell@prenhall.com or calling 1-201-236-7407.

For exploring properties on a sphere it is important that you have a model of a sphere that you can use. Some people find it helpful to purchase Lénárt Sphere® sets — a transparent sphere, a spherical compass, and a spherical "straight edge" that doubles as a protractor. They work well for small group explorations in the classroom and are available from Key Curriculum Press. However, considerably less expensive alternatives are available: A beach ball or basketball will work for classroom demonstrations, particularly if used with rubber bands large enough to form great circles on the ball. Students often find it convenient to use worn tennis balls ("worn" because the fuzz can get in the way) because they can be written on and are the right size for ordinary rubber bands to represent great circles. Also, many craft stores carry inexpensive plastic spheres that can be used successfully.

I strongly urge that you have a hyperbolic surface such as those described in Chapter 5. Unfortunately, such hyperbolic surfaces are not readily available commercially. However, directions for making such surfaces (out of paper or by crocheting) are contained in Chapter 5, and I will list patterns for making paper models and sources for crocheted hyperbolic surfaces at

www.math.cornell.edu/~dwh/eg00/supplements.html

as they become available. Most books that explore hyperbolic geometry do so by considering only one of the various "models" of hyperbolic geometry, which give representations of hyperbolic geometry in the same way that a map of a portion of the earth gives a representation of a portion of the earth. Each of these representations necessarily (see Chapter 16) distorts either straight lines or angles or both.

In addition, the use of dynamic geometry software such as *Geometers Sketchpad®*, *Cabri®*, or *Cinderella®* will enhance any geometry course. These software packages were originally written for exploring Euclidean plane geometry, but recent versions allow one to also dynamically explore spherical and hyperbolic geometries. I will maintain at the web address listed above links to information about these software packages and to web pages that give examples on how to use them for self-learning or in a classroom.

MY TEACHING BACKGROUND

My teaching is a product of Western Civilization. My known ancestors lived in England, Scotland, Ireland, Germany, and Luxembourg and I am a descendent from a long line of academics stretching back (according to family traditions) to at least the seventeenth century. My mode of teaching also has deep Western roots that reach back to the Socratic dialogues recorded by Plato in ancient Greece. More directly, my teaching has been influenced by my experiences in high school, college, and graduate school. In Ames, Iowa, my high school world literature teacher, Mary McNally, coaxed deep creative thinking out of us through her many writing assignments which she read with great interest in our ideas. At Swarthmore College in Pennsylvania instead of classes I spent my last two years in student participation seminars and tutorials, where I learned to take charge of my own learning and become an academic scholar. In graduate school at the University of Wisconsin my mentor, R.H. Bing, taught without lectures or textbooks in a style which is often known as the Moore Method, named after Bing's graduate mentor R.L. Moore at the University of Texas. (See [**TG**: Traylor] for more information on the Moore Method.) R.L. Moore received his PhD at the University of Chicago before the turn of the century and was one of the very first Americans to receive a PhD in mathematics in this country. My teaching of the geometry course and the writing of this book evolved from this background.

ACKNOWLEDGMENTS FOR FIRST EDITION

I acknowledge my debt to all the students and teachers who have attended my geometry courses. Most of these people have been students at Cornell or teachers in the surrounding area of upstate New York, but they also include students at Birzeit University in Palestine and teachers in the new South Africa. Without them this book would have been an impossibility.

Starting in 1986, Avery Solomon and I organized and taught a program of inservice courses for high school teachers under the financial support of Title IIA Grants administered by the New York State Department of Education. This is now called the Cornell/Schools Mathematics Resource Program (CSMRP). As a part of CSMRP we started recording classes and writing notes on the material. Some of the material in this book had its origins in those notes, but they never threatened to become a textbook. I thank Avery for his modeling of enthusiastic teaching, his sharp insights, and his insistence on preserving the teaching materials. In addition to Avery, my friends, Marwan Awartani, a professor at Birzeit University, and John Volmink, the director of the Centre for the Advancement of Science and Mathematics Education in Durban, South Africa, have for a long period of time consistently encouraged me to write this book.

A few years ago my colleague Maria Terrell suggested that five of us at Cornell who have been teaching non-traditional geometry courses (Avery Solomon, Bob Connelly, Tom Rishel, Maria, and I) submit a proposal to the National Science Foundation for a grant to write up materials on our courses. The fact that we were awarded the grant (in 1992) is largely due to Maria's persistence, clear thinking, and encouragement. It is this grant that gave me the necessary support to start the writing of this book. I thank the NSF's Program on Course and Curriculum Development for its support.

The major portions of this book were written during the 1992–93 academic year, in which I taught the course both semesters. Eduarda Moura was my teaching assistant for these courses. She was supported by the NSF grant to assist me by describing the classroom discussion and the student homework on which the content of this book is based. Much of this book (and especially the instructor's manual) are derived from her efforts. In addition to Eduarda, Kelly Gaddis, Beth Porter, Hal Schnee, and Justin Collins were also supported by the grant and made significant contributions to the writing of this book. I thank them all for their excellent contributions, their support of my work, and their friendship. The final writing and the decisions as to what to include and what not to include have all been mine, but they have been based on the foundation started with Avery and the CSMRP materials and continued with Eduarda, Justin, Kelly, Beth, and Hal during 1992–93.

Since the spring of 1992, the early drafts of the book have been used by me and others at Cornell and 13 other institutions. Various other individuals have worked through the book outside a classroom setting. From these students, instructors, and others I have received encouragement and much valuable feedback that has resulted in what I consider to be a better book. In particular, I want to thank the following persons for giving me feedback and ideas I have used in this final version: David Bray, Douglas Cashing, Helen Doerr, Jay Graening, Christine Kinsey, István Lénárt, Julie Lubell, Richard Pryor, Amanda Cramer and her students, Erica Flapan and her students, Linda Hill and her students, Tim Kurtz and his students, Judy Roitman and her students, Bob Strichartz and his students, and Walter Whitely and his students. Susan Alida spent many hours proofreading and refining the text, and was my consultant on matters of aesthetics.

ACKNOWLEDGMENTS FOR THIS EDITION

I wish to thank the instructors and students (all over the world) who have used the first edition: *Experiencing Geometry on Plane and Sphere.* Their responses were the first encouragement to expand and revise the book.

I wish to thank Jeffrey Weeks for introducing me to the current issues and upcoming experiments about the shape of our physical universe. He is the first to have informed me about the observations that are due to be performed in 2000–2001 that may allow a group of mathematicians and physicists (including Weeks) to determine the global geometry of our physical universe. It is my hope that this book provides the necessary background to understand these observations and determinations.

Without Daina Taimiņa's crocheted hyperbolic planes I would not have had the intuitive experiences that encouraged me to write this expansion and revision. The ideas for the expansion and revisions were discussed between us and much of the rewriting and expansion was completed with her able assistance.

In addition, the following persons gave me special comments that were incorporated into the expansion and revision: sarah-marie belcastro (University of Northern Iowa), David Bellamy (University of Delaware), Gian Mario Besana (Eastern Michigan University), Alexander Bogomolny (CTK Software, East Brunswick, NJ), Katherine Borgen (University of British Columbia), Sean Bradley (Clarke College), David Dennis (University of Texas at El Paso), Kelly Gaddis (Lewis and Clarke University), Paul J. Gies (University of Maine at Farmington), Chaim Goodman-Strauss (University of Arkansas), Alice Guckin (College of Saint Scholastica), Cathy Hayes, Keith Henderson (Thomas Jefferson School, St. Louis, MO), George H. Litman (National-Louis University), Jane-Jane Lo (Ithaca College and Cornell University), Alan Macdonald (Luther College), John McCleary (Vassar College), Nathaniel Miller (Cornell University), David Mond (University of Warwick), Colm Mulcahy (Spelman College), Jodie Novak (University of Northern Colorado), David A. Olson (MTU), Mary Platt (Salem State College, MA), John Poland (Carleton College), Nancy Rodgers (Hanover College), Thomas Sibley (St. John's University), Judith Roitman (University of Kansas), Frances Rosamond (National University), Avery Solomon (Cornell University), Daniel H. Steinberg (Case Western Reserve University), Robert Stolz (University of the Virgin Islands), John Sullivan (University of Illinois – Urbana), Margaret Symington (University of Texas), George Tintera (Texas A&M University – Corpus Christi), Susan Tolman (University of Illinois – Urbana), Andy Vidan (Cornell University), Tad Watanabe (Towson State University), Walter Whiteley (York University), Jeffrey Weeks (Canton, NY), Steve Weissburg (Ithaca High School, Ithaca, NY), Cindy Wyels (California Lutheran University), and Michelle Zandieh (Arizona State University). I may have inadvertently left out a few names; if so, I apologize.

I produced the entire manuscript (typing, formatting, drawings, and final layout) using the integrated word processing software WordPro. Finally, I wish to thank George Lobell, Senior Editor at Prentice Hall, for his encouragement and support and for the vision and enthusiasm with which he shepherded both editions through the publication process. I also thank Betsy Williams, Production Editor, for thoughtfully improving the style of this book, which has made the book much more pleasing and readable.

David W. Henderson
Ithaca, NY

HOW TO USE THIS BOOK

Do not just pay attention to the words;
Instead pay attention to meanings behind the words.
But, do not just pay attention to meanings behind the words;
Instead pay attention to your deep experience of those meanings.
— Tenzin Gyatso, The Fourteenth Dalai Lama[†]

This quote expresses the philosophy on which this book is based. Most of the chapters start intuitively so that they are accessible to a general reader with no particular mathematics background except imagination and a willingness to struggle with one's own experience of the meanings. However, the discussions in the book were written for mathematics majors and mathematics teachers and thus assume of the reader a corresponding level of interest and mathematical sophistication.

This book will present you with a series of problems. You should explore each question and write out your thinking in a way that can be shared with others. By doing this you will be able to actively develop ideas prior to passively reading or listening to comments of others. When working on the problems, you should be open-minded and flexible and let your thinking wander. Some problems will have short, fairly definitive answers, and others will lead into deep areas of meaning that can be probed almost indefinitely. You should not accept anything just because you remember it from school or because some authority says it's good. Insist on understanding (or seeing) why it is true or what it means for you. Pay attention to *your* deep experience of these meanings.

You should think about each problem and express your thinking even when you know you cannot complete it. This is important because

---

[†]From an unpublished lecture in London, April 1984. Used here by permission.

- It helps build self confidence.
- You will see what your real difficulties are.
- When you see a solution or proof later, you will more likely see it as answering a question that you have.

An important thing to keep in mind is that there is no one correct solution. There are many different ways of solving the problems — as many as there are ways of understanding the problems. *Insist on understanding* (or seeing) why it is true or what it means to you. Everyone understands things in a different way, and one person's "obvious" solution may not work for you. However, it is helpful to talk with others — listen to their ideas and confusions and then share your ideas and confusions with them. In the experience of those using this book and the earlier edition, it is very important to be able to talk with others in small groups whether inside or outside of class. In fact, small groups have successfully gone through this book as self-study without a teacher.

Also, some of the problems are difficult to visualize in your head. Make models, draw pictures, use rubber bands on a ball, use scissors and paper — play!

For exploring properties on a sphere it is important that you have a model of a sphere that you can use. You can draw on worn tennis balls ("worn" because the fuzz can get in the way) and they are the right size for ordinary rubber bands to represent great circles. You may find useful clear plastic spheres from craft stores. Most any ball you have around will work — you can even use an orange and then eat it when you get hungry!

For exploring the geometric properties of a hyperbolic plane it is very important to have a hyperbolic surface in your hands. Instructions on how you can make (either out of paper or by crocheting) hyperbolic surfaces are contained in the beginning of Chapter 5. It will be very helpful to your understanding of the hyperbolic plane for you to actually make one of these hyperbolic surfaces yourself.

How I Use This Book in a Course

In my course *the distinction between learning activities and assessment activities is blurred.* I present a sequence of problems (together with motivation, discussion of contexts, and connections of the problems with other areas of mathematics and life). I tell the students

Write out your thinking to each problem. We will return your papers with comments about your solutions. Respond to our comments — use them as invitations to explore, to clarify your understanding of the problem, or to clarify our understanding of your solution. Keep responding until you understand. Turn in whatever your thinking is on a question even if only to say "I don't understand such and such" or "I'm stuck here"; be as specific as possible. Feel free to ask questions. This will allow the sharing of ideas, and you will benefit more from class sessions.

The students then work on the problems either individually or in small groups and report their thinking back to me and the class. This cycle of **writing, comments, discussion** continues on each problem until both the students and I are satisfied, unless external constraints of time and resources intervene.

Some problems are investigated by my students in small cooperative learning groups (with no written work) and the groups report back to the class, or not, depending on how it goes.

What I have discovered is that in this process not only have the students learned from the course, but also I have learned much about geometry from them. At first I was surprised; how could I, the teacher, learn mathematics from the students? But this learning has continued for 25 years and I now expect its occurrence. In fact, as I expect it more and more and learn to listen more effectively to them, I find that a greater portion of my students show something new to me about geometry. For more discussion of this, see the "Message to the Reader" on pages xxviii–xxxi.

For a final project I usually ask each student to pick a chapter that we have not covered in class and to investigate it on his or her own (with my assistance as necessary in office hours). I have found from the experiences of my students that all the chapters work for such projects.

BUT, DO IT YOUR OWN WAY

From feedback I have received from instructors using the first edition of this book and preliminary versions of this edition, I have learned that many of the most successful uses of this book in the classroom are when the instructors does not follow this book! In many of these successful courses, the instructors have used some of the problems from this book and then added their own favorite problems. The chapters in the book have been used in a variety of different orders and at many different speeds. Some instructors have, with careful organization and direction, successfully covered the whole of the first edition of this book in a semester. Other instructors have let their students explore and wander in such a way that they covered less than half of the book. Some instructors use the book in a course with traditional preliminary and final exams. Others have had no exams, but instead relied on portfolios, student journals, and/or projects. Some have used small cooperative learning groups regularly throughout the semester and others have never divided the students into small groups.

Also, the book has been used in courses for sophomores who are prospective teachers — either elementary or secondary. Or, in courses for senior mathematics majors. Or, in courses for Masters students in Education. Many of the problems in this book (but not the book itself) have been used successfully with freshman-level courses for students weak in mathematics ("liberal arts mathematics"). Portions of the book have also been used successfully in a freshman writing course.

CHAPTER SEQUENCES

I have attempted to make the chapters more independent of each other than was the case in the first edition. But there is a minimum core that is necessary before exploring the other chapters.

Minimum core: Cover, at a minimum, Chapters 1–3, 6, 8, and either 7 or 10.

All chapters after Chapter 10 are independent: After exploring the minimum core, all of the other chapters are essentially independent. Each chapter refers to some results from other chapters but these references can be looked up without destroying the experience of the chapter you are reading.

Starred problems and sections: Certain problems and sections in this book require from the reader a background more advanced than first-semester calculus — these sections are indicated with an asterisk (*) and the background required is indicated (usually at the beginning of the chapter).

Here are some chapter sequences that put emphasis on different aspects of geometry:

To use this book the same as the first edition: Leave out Chapters 5, 14, 16, 17, and 22; and ignore the references to hyperbolic geometry and the hyperbolic plane in the remaining chapters.

To emphasize traditional topics in Euclidean plane geometry: Explore Chapters 1–3, (4), 6, 8–14, 18, 21. This includes much geometry on spheres (and cones and cylinders), which is needed to bring out the meanings of Euclidean geometry.

To emphasize spherical geometry: Explore Chapters 1–3, 6–9, 11, 12, 15, 19.

To emphasize hyperbolic geometry: Explore hyperbolic plane parts of Chapters 1–3, 5–12, 14, and 16.

To get to Chapter 22 on the shape of space as quickly as possible: Explore Chapters 1–8, 17, and Problems **20.1** and **20.6**.

For chapters using similar triangles (Chapters 14–19): Explore Problem **12.1** and **13.3** or assume the two Criteria for Similar Triangles (Problem **13.3**).

For three-dimensional investigations in Chapters 21 and 22: At least explore Problem **20.1** and assume the results of Problem **20.6**.

These chapter sequences are summarized in the following table:

Goal	Chapters																
Minimum Core	1–3, 6, 8, and either 7 or 10																
Use book similar to first edition	4		7	9	10	11	12	13		15			18	19	20	21	
To emphasize Euclidean geometry	-4			9	10	11	12	13	14				18			21	
To emphasize spherical geometry			7	9		11	12			15				19			
To emphasize hyperbolic geometry		5	7	9	10	11	12		14		16						
To get to Chapter 22 "Shape of Space" as soon as possible	4	5	7									17			20.1 & 20.6		22
To prove and apply similar triangle criteria (**13.3**)				9	10		12	13	14	15	16	17	18	19			

MESSAGE TO THE READER

In mathematics, as in any scientific research, we find two tendencies present. On the one hand, the tendency toward *abstraction* seeks to crystallize the *logical* relations inherent in the maze of material that is being studied, and to correlate the material in a systematic and orderly manner. On the other hand, the tendency toward *intuitive understanding* fosters a more immediate grasp of the objects one studies, a live *rapport* with them, so to speak, which stresses the concrete meaning of their relations.

As to geometry, in particular, the abstract tendency has here led to the magnificent systematic theories of Algebraic Geometry, of Riemannian Geometry, and of Topology; these theories make extensive use of abstract reasoning and symbolic calculation in the sense of algebra. Notwithstanding this, it is still as true today as it ever was that intuitive understanding plays a major role in geometry. And such concrete intuition is of great value not only for the research worker, but also for anyone who wishes to study and appreciate the results of research in geometry.

— David Hilbert[**SE:** Hilbert, p. iii]

These words were written in 1934 by the "father of Formalism" David Hilbert in the Preface to *Geometry and the Imagination* by Hilbert and S. Cohn-Vossen. Hilbert has emphasized the point I wish to make in this book:

Meaning is important in mathematics and geometry is an important source of that meaning.

I believe that mathematics is a natural and deep part of human experience and that experiences of meaning in mathematics are accessible to everyone. Much of mathematics is not accessible through formal approaches except to those with specialized learning. However, through the use of non-formal experience and geometric imagery, many levels of meaning in mathematics can be opened up in a way that most human beings can experience and find intellectually challenging and stimulating.

Formalism contains the power of the meaning but not the meaning. It is necessary to bring the power back to the meaning.

A proof as we normally conceive of it is not the goal of mathematics — it is a tool — a means to an end. The goal is understanding. Without understanding we will never be satisfied — with understanding we want to expand that understanding and to communicate it to others.

Many formal aspects of mathematics have now been mechanized and this mechanization is widely available on personal computers or even handheld calculators, but the experience of meaning in mathematics is still a human enterprise that is necessary for creative work.

In this book I invite the reader to explore the basic ideas of geometry from a more mature standpoint. I will suggest some of the deeper meanings, larger contexts, and interrelations of the ideas. I am interested in conveying a different approach to mathematics, stimulating the reader to take a broader and deeper view of mathematics, and to experience for her- or himself a sense of mathematizing. Through an active participation with these ideas, including exploring and writing about them, people can gain a broader context and experience. This active participation is vital for anyone who wishes to understand mathematics at a deeper level, or anyone wishing to understand something in their experience through the vehicle of mathematics.

This is particularly true for teachers or prospective teachers who are approaching related topics in the school curriculum. All too often we convey to students that mathematics is a closed system, with a single answer or approach to every problem, and often without a larger context. I believe that even where there are strict curricular constraints, there is room to change the meaning and the experience of mathematics in the classroom.

PROOF AS CONVINCING ARGUMENT
THAT ANSWERS — WHY?

Much of our view of the nature of mathematics is intertwined with our notion of what is a proof. This is often particularly true with geometry, which has traditionally been taught in high school in the context of "two-column" proofs. The course materials in this book are based on a view of proof as a convincing argument that answers a why-question.

Why is $3 \times 2 = 2 \times 3$? To say, "It follows from the Commutative Law" does not answer the why-question. But most people will be convinced by, "I can count three 2's and then two 3's and see that they are both equal to the same six." OK, now why is $2{,}657{,}873 \times 92{,}564 = 92{,}564 \times 2{,}657{,}873$? We cannot count this — it is too large. But is there a way to see $3 \times 2 = 2 \times 3$ without counting? Yes.

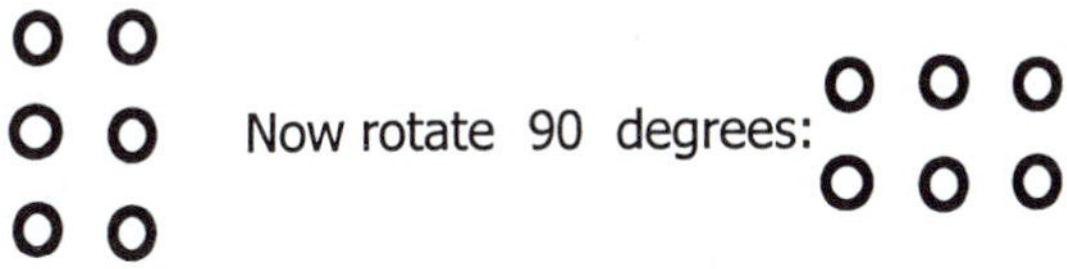

Figure 0.1   Why is $3 \times 2 = 2 \times 3$?

Most people will not have trouble extending this proof to include $2{,}657{,}873 \times 92{,}564 = 92{,}564 \times 2{,}657{,}873$ or the more general $n \times m = m \times n$. Note that for the above to make sense I must have a meaning for 3×2 and a meaning for 2×3 *and these meanings must be different.* So naturally I have the question: "Why (or in what sense) are these meanings related?" A proof should help me experience relationships between the meanings. In my experience, to perform the formal mathematical induction proof starting from Peano's Axioms does not answer anyone's why-question unless it is such a question as: "Why does the Commutative Law follow from Peano's Axioms?" Most people (other than logicians) have little interest in that question.

> **CONCLUSION 1:** *In order for me to be satisfied by a proof, the proof must answer my why-question and relate my meanings of the concepts involved.*

As further evidence toward this conclusion, you have probably had the experience of reading a proof and following each step logically but still not being satisfied because the proof did not lead you to experience the answer to your why-question. In fact most proofs in the literature are not written out in such a way that it is possible to follow each step in a logical formal way. Even if they were so written, most proofs would be too long and complicated for a person to check each step. Furthermore, even among mathematics researchers, a formal logical proof that they can follow step-by-step is not always satisfying. For example, my short-est research paper ["A simplicial complex whose product with any ANR is a simplicial complex," *General Topology* **3** (1973), pp. 81–83] has a very concise simple proof that anyone who understands the terms involved can easily follow logically step-by-step. But, I have received more questions from other mathematicians about that paper than about any of my other research papers and most of the questions were of the sort: "Why is it true?" "Where did it come from?" "How did you see it?" They accepted the proof logically but were not satisfied.

Let us look at another example — the Vertical Angle Theorem: *If l and l' are straight lines, then the angle α is congruent to the angle β.*

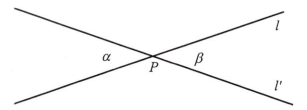

Figure 0.2   Vertical Angle Theorem

As we will see in Problems **3.1** and **3.2** of this book, the proof of this theorem that is most convincing to someone depends on the mean-ings that person has of "angle" and "congruence." Some years ago, after I had been teaching this geometry course for over ten years, several proofs that were convincing to me were presented by students in the class. But one student found the proofs not so convincing and offered a very straightforward simple proof of her own. My first reaction was that her argument could not possibly be a proof — it was too simple and did not involve everything in the standard proof. But she persisted patiently for several days and my meanings for angle congruence deepened. Now her proof is much more convincing to me than the standard one. I hope you will have similar experiences while working through this book.

CONCLUSION 2: *A proof that satisfies someone else may not satisfy me because their meanings and why-questions are different from mine.*

You may ask, "But, at least in plane geometry, isn't an angle an angle? Don't we all agree on what an angle is?" Well, yes and no. Consider this acute angle:

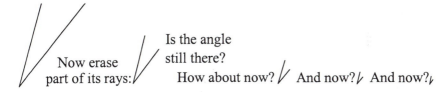

Now erase part of its rays: Is the angle still there? How about now? And now? And now?

Figure 0.3   Where is the angle?

The angle is somehow *at the corner*, yet it is difficult to express this formally. As evidence, I looked in all the plane geometry books in the university library and found their definitions for "angle." I found nine different definitions! Each expressed a different meaning or aspect of "angle" and thus, each would potentially lead to a different proof of the Vertical Angle Theorem. We will see this more when we discuss Problems **3.1** and **3.2**.

Sometimes we have legitimate why-questions even with respect to statements traditionally accepted as axioms. The Commutative Law above is one possible example. Another one is Side-Angle-Side (or SAS): *If two triangles have two sides and the included angle of one congruent to two sides and the included angle of the other, then the triangles are congruent.* You can find SAS listed in some geometry textbooks as an axiom to be assumed; in others it is listed as a theorem to be proved and in still others as a definition of the congruency of two triangles. But clearly one can ask, "Why is SAS true in the plane?" This is especially true because SAS is false for (geodesic) triangles on the sphere. So one can naturally ask, "Why is SAS true on the plane but not on the sphere?"

I have been teaching a geometry course based on the material in this book for a long time now (since 1974). One might expect that I have seen everything. But every year, about one-third of the students will show me a meaning or way of looking at the geometry that I have never thought of before and thus my own meaning and experience of geometry deepen. Looking back, I notice that these students who have shown me something new are mostly persons whose cultural backgrounds or race or gender are different from mine; and this is true even though most of the students in the class and I are white males. (For details of this data and further discussion see my article "I Learn Mathematics From My Students — Multiculturalism in Action", *For the Learning of Mathematics* **16** (1996), pp. 34–40, or on my webpage.)

CONCLUSION 3: *Persons who differ in terms of cultural background, race, gender are likely to have different meanings and thus have different why-questions and different proofs.*

You should check this out in your own experience. We should listen carefully to meanings and proofs expressed by all persons. We should also be more critical of many of the standard histories of mathematics and mathematicians, which have a decidedly Eurocentric emphasis.

As we personally experience Conclusions 1, 2, 3 above, we are led to the following conclusion.

CONCLUSION 4: *If I experience 1, 2, and 3, then other persons (for example, my students) are also likely to have similar experiences.*

Chapter 1

WHAT IS STRAIGHT?

Straight is that of which the middle is in front of both extremities.

— Plato, *Parmenides,* 137 E [**AT**: Plato]

A *straight line* is a line that lies symmetrically with the points on itself.

— Euclid, *Elements*, Definition 4 [Appendix A]

Wisdom will save you from the ways of wicked men, from men whose words are perverse, who leave the straight paths to walk in the dark ways, ...whose paths are crooked and who are devious in their ways.

— *The Holy Bible*, Proverbs 2:12–15

Verily, this is My Way, leading straight, follow it: Follow not other paths: They will scatter you about from His great path.

— *Holy Qur'an*, Sura VI, verse 153

In keeping with the spirit of the approach to geometry discussed in the *Preface* and *Message to the Reader*, we begin with a question that encourages you to explore deeply a concept that is fundamental to all that will follow. We ask you to build a notion of straightness for yourself rather than accept a certain number of assumptions about straightness. Though it is difficult to formalize, straightness is a natural human concept.

1

PROBLEM 1.1   WHEN DO YOU CALL A LINE STRAIGHT?

Look to your experiences. It might help to think about how you would explain straightness to a 5-year-old (or how the 5-year-old might explain it to you!). If you use a "ruler," how do you know if the ruler is straight? How can you check it? What properties do straight lines have that distinguish them from non-straight lines?

Think about the question in four related ways:

a. *How can you check in a practical way if something is straight — without assuming that you have a ruler, for then we will ask, "How can you check that the ruler is straight?"*

b. *How do you construct something straight — lay out fence posts in a straight line, or draw a straight line?*

c. *What symmetries does a straight line have?*

d. *Can you write a definition of "straight line"?*

SUGGESTIONS

Look at your experience. At first, you will look for examples of physical world or natural straightness. Go out and actually try walking along a straight line and then along a curved path; try drawing a straight line and checking that a line already drawn is straight.

Look for things that you call "straight." Where do you see straight lines? Why do you say they are straight? Look for both physical lines and nonphysical uses of the word "straight."

You are likely to bring up many ideas of straightness. It is necessary then to think about what is common among all of these straight phenomena.

As you look for properties of straight lines that distinguish them from non-straight lines, you will probably remember the following statement (which is often taken as a definition in high school geometry): *A straight line is the shortest distance between two points.* But can you ever measure the lengths of all the paths between two points? How do you find the shortest path? If the shortest path between two points is in fact a straight line, then is the converse true? Is a straight line between two points always the shortest path? We will return to these questions in later chapters.

A powerful approach to this problem is to think about lines in terms of symmetry. This will become increasingly important as we go on to other surfaces (spheres, cones, cylinders, etc.) Two symmetries of lines are:

- Reflection symmetry in the line, also called bilateral symmetry — reflecting (or mirroring) an object over the line.

Figure 1.1   Reflection symmetry of a straight line

- Half-turn symmetry — rotating 180° about any point on the line.

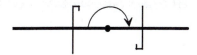

Figure 1.2   Half-turn symmetry of a straight line

Although we are focusing on a symmetry of the line in each of these examples, notice that the symmetry is not a property of the line by itself but includes the line and the space around the line. The symmetries preserve the local environment of the line. Notice how in reflection and half-turn symmetry the line and its local environment are both part of the symmetry action and how the relationship between them is integral to the action. In fact, reflection in the line does not move the line at all but exhibits a way in which the spaces on the two sides of the line are the same.

DEFINITIONS. An *isometry* is a transformation of a region of space that preserves distances and angle measures. A *symmetry of a figure* is an isometry of a region of space that takes the figure (or the portion of it in the region) onto itself. You will show in Problem **8.3** that every isometry of the plane is either a translation, a rotation, a reflection, or a composition of them.

Try to think of other symmetries of a line as well (there are quite a few). Some symmetries hold only for straight lines, while some work with curves too. Try to determine which ones are specific to straight lines and why. Also think of practical applications of these symmetries for constructing a straight line or for determining if a line is straight.

How Do You Construct a Straight Line?

Imagine (or actually try!) walking while pulling a long silk thread with a small stone attached. When will the stone follow along your path? Why? This property is used to pick up a fallen water skier. The boat travels by the skier along a straight line and thus the tow rope follows the path of the boat. Then the boat turns in an arc in front of the skier. Because the boat is no longer following a straight path, the tow rope moves in toward the fallen skier.

Another idea to keep in mind is that straightness must be thought of as a local property. Part of a line can be straight even though the whole line may not be. For example, if we agree that this line is straight,

and then we add a squiggly part on the end, like this,

would we now say that the original part of the line is not straight, even though it hasn't changed, only been added to? Also note that we are not making any distinction here between "line" and "line segment." The more generic term "line" generally works well to refer to any and all lines and line segments, both straight and non-straight.

As for how to construct a straight line, one method is simply to fold a piece of paper; the edges of the paper needn't even be straight. This utilizes symmetry (can you see which one?) to produce the straight line. Carpenters also use symmetry to determine straightness — they put two boards face to face, plane the edges until they look straight, and then turn one board over so the planed edges are touching. See Figure 1.3. They then hold the boards up to the light. If the edges are not straight, there will be gaps between the boards where light will shine through.

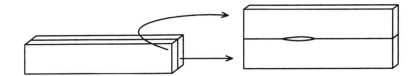

Figure 1.3 Carpenter's method for checking straightness

When grinding an extremely accurate flat mirror, the following technique is sometimes used: Take three approximately flat pieces of glass and put pumice between the first and second pieces and grind them together. Then do the same for the second and the third pieces and then for the third and first pieces. Repeat many times and all three pieces of glass will become very accurately flat. See Figure 1.4. Do you see why this works? What does this have to do with straightness?

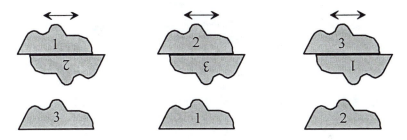

Figure 1.4   Grinding flat mirrors

What symmetries does a straight line have? How do they fit with the examples that you have found and those mentioned above? Can we use any of the symmetries of a line to define straightness?

The intersection of two (flat) planes is a straight line — why does this work? Does it help us understand "straightness"?

Think about and formulate some answers for these questions before you read any further. You are the one laying down the definitions. Do not take anything for granted unless you see why it is true. No answers are predetermined. You may come up with something that we have never imagined. Consequently, it is important that you persist in following your own ideas. Reread the section "How to Use This Book" at the beginning of this book.

You should not read further until you have expressed your thinking and ideas through writing or talking to someone else.

THE SYMMETRIES OF A LINE

Reflection-in-the-line symmetry: It is most useful to think of reflection as a "mirror" action with the line as an axis rather than as a "flip-over" action which involves an action in 3-space. In this way one can extend the notion of reflection symmetry to a sphere (the flip-over action is not possible on a sphere). Notice that this symmetry cannot be used as a definition for straightness because we use straightness to define reflection symmetry. This same comment applies to most of the other symmetries discussed below.

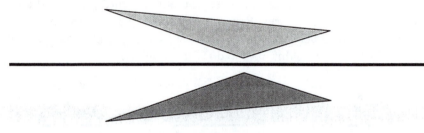

Figure 1.5   Reflection-in-the-line symmetry

In Figures 1.5–1.11, the light gray triangle is the image of the dark gray triangle under the action of the symmetry on the space around the line.

- *Practical application*: One can produce a straight line by folding a piece of paper because this action forces symmetry along the crease. Above we showed a carpenter's example.

Reflection-perpendicular-to-the-line symmetry: A reflection through **any** axis perpendicular to the line will take the line onto itself. Note that circles also have this symmetry about any diameter. See Figure 1.6.

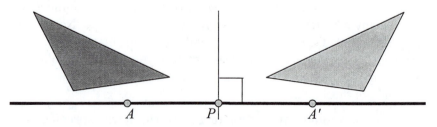

Figure 1.6   Reflection-perpendicular-to-the-line symmetry

♦ *Practical applications*: You call tell if a straight segment is perpendicular to a mirror by seeing if it looks straight with its reflection. Also, a straight line can be folded onto itself.

Half-turn symmetry: A rotation through half of a full revolution about any point P on the line takes the part of the line before P onto the part of the line after P and vice versa. Note that some non-straight lines, such as the letter "Z" also have half-turn symmetry.

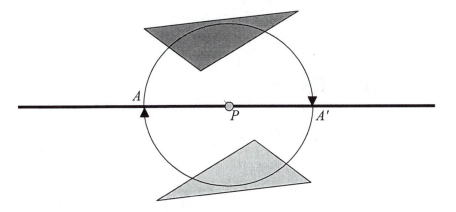

Figure 1.7   Half-turn symmetry

♦ *Practical applications*: Half-turn symmetry exists for the slot on a screw and the tip of the screwdriver (unless you are using Phillips-head screws and screwdrivers, which have quarter-turn symmetry) and thus one can more easily put the tip of the screwdriver into the slot.

Rigid-motion-along-itself symmetry: For straight lines in the plane, we call this *translation symmetry*. Any portion of a straight line may be moved along the line without leaving the line. This property of being able to move rigidly along itself is not unique to straight lines; circles (rotation symmetry) and circular helixes (screw symmetry) have this property as well. See Figure 1.8.

♦ *Practical applications*: Slide joints such as in trombones, drawers, nuts and bolts, pulleys, etc. all utilize this symmetry.

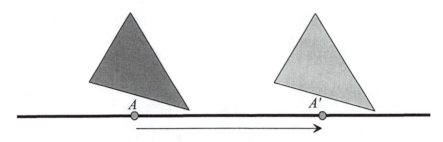

Figure 1.8   Rigid-motion-along-itself symmetry

3-dimensional-rotation symmetry: In a 3-dimensional space, rotate the line around itself through *any* angle using itself as an axis.

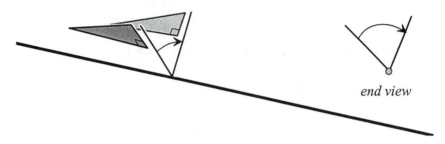

end view

Figure 1.9   3-dimensional-rotation symmetry

- *Practical applications*: This symmetry can be used to check the straightness of any long thin object such as a stick by twirling the stick with itself as the axis. If the stick does not appear to wobble, then it is straight. This is used for pool cues, axles, hinge pins, and so forth.

Central symmetry, or point symmetry: Central symmetry through the point *P* sends any point *A* to the point on the line determined by *A* and *P* at the same distance from *P* but on the opposite side from *P*. In two dimensions central symmetry does not differ from half-turn symmetry in its end result, but they do differ in the ways we imagine them and construct them.

- In 3-space, central symmetry produces a result different than any single rotation or reflection (though one can check that it does give the same result as the composition of three reflections through mutually perpendicular planes). To experience central symmetry in 3-space, hold your hands in front of you with the palms facing each other and your left thumb up and your right thumb down. Your two hands now have approximate central symmetry about a point midway between the center of the palms and this symmetry cannot not be produced by any reflection or rotation.

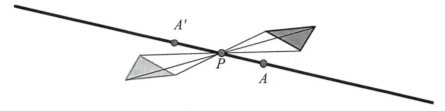

Figure 1.10   Central symmetry

Similarity or self-similarity "quasi-symmetry": Any segment of a straight line (and its environs) is similar to (that is, can be magnified or shrunk to become the same as) any other segment. This is not a symmetry because it does not preserve distances but it could be called a "quasi-symmetry" because it *does* preserve the measure of angles.

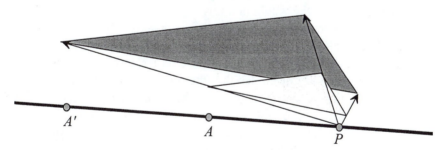

Figure 1.11   Similarity "quasi-symmetry"

- Logarithmic spirals such as the chambered nautilus have self-similarity as do many fractals. (See example in Figure 1.12.)

Clearly, other objects besides lines have some of the symmetries mentioned here. It is important for you to construct your own such examples and attempt to find an object that has all of the symmetries but is not a line. This will help you to experience that straightness and the seven symmetries discussed here are intimately related. You should come to the conclusion that while other curves and figures have some of these symmetries, only the straight line has all of them.

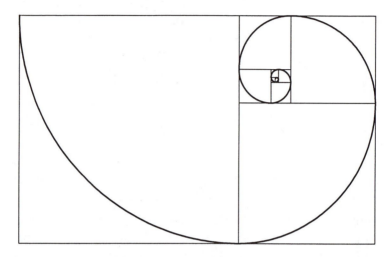

Figure 1.12   Logarithm spiral

Returning to one of the original questions, how would we construct a straight line? One way would be to use a "straight edge" — something that we accept as straight. Notice that this is different from the way that we would draw a circle. When using a compass to draw a circle, we are not starting with a figure that we accept as circular; instead, we are using a fundamental property of circles that the points on a circle are a fixed distance from the center. Can we use the symmetry properties of a straight line to construct a straight line? Remember the examples earlier in this chapter. Is there a tool (serving the role of a compass) that will draw a straight line? For an interesting discussion of this question see *How to Draw a Straight Line: A Lecture on Linkages* by A.B. Kempe [**Z**: Kempe], which shows the apparatus depicted in Figure 1.13.

The links labeled with the same letter must have the same length. The fact that this apparatus draws a straight line is the subject of Problem **11.3b**. See [**SE**: Hilbert, pp. 272–73] for another discussion of this topic. The discovery of this linkage about 1870 is variously attributed to the French army officer, Charles-Nicolas Peaucellier (1832-1913), and to Lippman Lipkin, who lived in Lithuania and studied in Saint Petersburg. (See also Phillip Davis' delightful little book *The Thread* [**Z**: Davis], Chapter IV.)

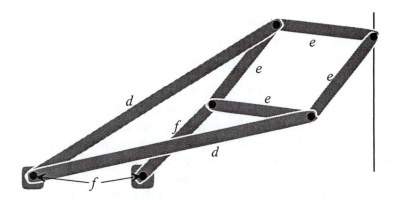

Figure 1.13   Apparatus for drawing a straight line

LOCAL (AND INFINITESIMAL) STRAIGHTNESS

Previously, you saw how a straight line has reflection-in-the-line symmetry and half-turn symmetry: One side of the line is the same as the other. But, as pointed out above, straightness is a local property in that whether a segment of a line is straight depends only on what is near the segment and does not depend on anything happening away from the line. Thus each of the symmetries must be able to be thought of (and experienced) as applying only locally. This will become particularly important later when we investigate straightness on the cone and cylinder. (See the discussions in Chapter 4.) For now, it can be experienced in the following way:

> *When a piece of paper is folded not in the center, the crease is still straight even though the two sides of the crease on the paper are not the same. (See Figure 1.14.)*

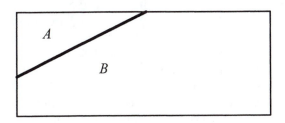

Figure 1.14   Reflection symmetry is local

So, what is the role of the sides when we are checking for straightness using reflection symmetry? Think about what is important near the crease in order to have reflection symmetry.

When we talk about straightness as a local property, you may bring out some notions of scale. For example, if one sees only a small portion of a very large circle, it will be indistinguishable from a straight line. This can be experienced easily on many of the modern graphing programs for computers. Also a microscope with a zoom lens will provide an experience of zooming. If a curve is smooth (or differentiable), then if one "zooms in" on any point of the curve, eventually the curve will be indistinguishable from a straight line segment. See Figure 1.15.

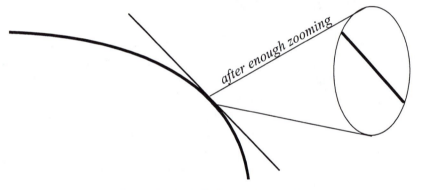

Figure 1.15   Infinitesimally straight

We sometimes use the terminology, *infinitesimally straight*, in place of the more standard terminology, *differentiable*. We say that a curve is *infinitesimally straight* at a point *p* if there is a straight line *l* such that if we zoom in enough on *p*, the line and the curve become indistinguishable[†]. When the curve is parametrized by arc length this is

---

[†]This is equivalent to the usual definitions of being differentiable at *p*. For example, if $t(x) = f(p) + f'(p)(x - p)$ is the equation of the line tangent to the curve $(x, f(x))$ at the point $(p, f(p))$, then, given $\varepsilon > 0$ (the distance of indistinguishability), there is a $\delta > 0$ (the radius of the zoom window) such that, for $|x - p| < \delta$ (for *x* within the zoom window), $|f(x) - t(x)| < \varepsilon$ [$f(x)$ is indistinguishable from $t(x)$]. This last inequality may look more familiar in the form:

$$f(x) - t(x) = f(x) - f(p) - f'(p)(x - p) = \{ [f(x) - f(p)]/(x - p) - f'(p) \}(x - p) < \varepsilon.$$

In general, the value of δ might depend on *p* as well as on ε. Often the term *smooth*

equivalent to the curve having a well-defined velocity vector at each point.

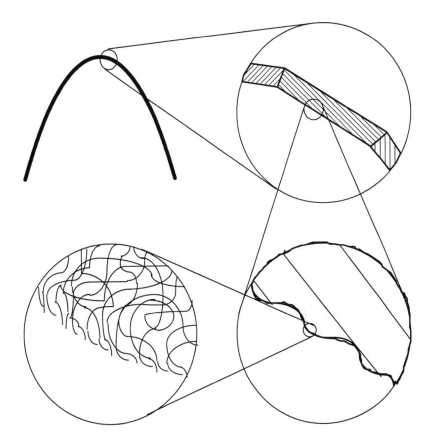

Figure 1.16   Straightness and smoothness depend on the scale

In contrast, we can say that a curve is *locally straight at a point* if that point has a neighborhood that is straight. In the physical world the usual use of both *smooth* and *locally straight* is dependent on the scale at which they are viewed. For example, we may look at an arch made out of wood — at a distance it appears as a smooth curve; then as we move in closer we see that the curve is made by many short straight pieces of

---

is used to mean continuously differentiable, which the interested reader can check is equivalent (on closed finite intervals) to, for each $\varepsilon > 0$, there being one $\delta > 0$ that works for all p.

finished (planed) boards, but when we are close enough to touch it, we see that its surface is made up of smooth waves or ripples, and under a microscope we see the non-smoothness of numerous twisting fibers. See Figure 1.16.

Chapter 2

STRAIGHTNESS ON SPHERES

... it will readily be seen how much space lies between the two places themselves on the circumference of the large circle which is drawn through them around the earth. ... [W]e grant that it has been demonstrated by mathematics that the surface of the land and water is in its entirety a sphere, ... and that any plane which passes through the center makes at its surface, that is, at the surface of the earth and of the sky, great circles, and that the angles of the planes, which angles are at the center, cut the circumferences of the circles which they intercept proportionately, ...
— Ptolemy, *Geographia* (ca. 150 AD) Book One, Chapter II

Drawing on the understandings about straightness you developed in Problem **1.1**, the second problem asks you to investigate the notion of straightness on a sphere. It is important for you to realize that, if you are not building a notion of straightness for yourself (for example, if you are taking ideas from books without thinking deeply about them), then you will have difficulty building a concept of straightness on surfaces other than a plane. Only by developing a personal meaning of straightness for yourself does it become part of your active intuition. We say *active* intuition to emphasize that intuition is in a process of constant change and enrichment, that it is not static.

PROBLEM 2.1 WHAT IS STRAIGHT ON A SPHERE?

a. *Imagine yourself to be a bug crawling around on a sphere. (This bug can neither fly nor burrow into the sphere.) The bug's universe is just the surface; it never leaves it. What is "straight" for this bug? What will the bug see or experience as straight? How can you convince yourself of this? Use the properties of*

15

straightness (such as symmetries) that you talked about in Problem **1.1**.

b. *Show (that is, convince yourself, and give an argument to convince others) that the great circles on a sphere are straight with respect to the sphere, and that no other circles on the sphere are straight with respect to the sphere.*

SUGGESTIONS

Great circles are those circles which are the intersection of the sphere with a plane through the center of the sphere. Examples include any longitude line and the equator on the earth. Note that any pair of opposite points can be considered as the poles, and thus the equator and longitudes with respect to any pair of opposite points will be great circles. See Figure 2.1.

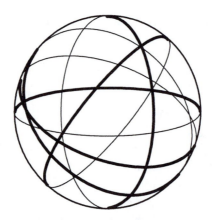

Figure 2.1   Great circles

The first step to understanding this problem is to convince yourself that great circles are straight lines on a sphere. Think what it is about the great circles that would make the bug experience them as straight. To better visualize what is happening on a sphere (or any other surface, for that matter), **you must use models**. This is a point we cannot stress enough. The use of models will become increasingly important in later problems, especially those involving more than one line. You must make lines on a sphere to fully understand what is straight and why. An orange or an old, worn tennis ball work well as spheres, and rubber bands make good lines. Also, you can use ribbon or strips of paper. Try placing these items on the sphere along different curves to see what happens.

Also look at the symmetries from Problem **1.1** to see if they hold for straight lines on the sphere. The important thing here is to **think in terms of the surface of the sphere, not the solid 3-dimensional ball**. Always try to imagine how things would look from the bug's point of view. A good example of how this type of thinking works is to look at an insect called a water strider. The water strider walks on the surface of a pond and has a very 2-dimensional perception of the world around it — to the water strider, there is no up or down; its whole world consists of the 2-dimensional plane of the water. The water strider is very sensitive to motion and vibration on the water's surface, but it can be approached from above or below without its knowledge. Hungry birds and fish take advantage of this fact. This is the type of thinking needed to adequately visualize properties of straight lines on the sphere. For more discussion of water striders and other animals with their own varieties of intrinsic observations, see the delightful book, *The View from the Oak*, by Judith and Herbert Kohl [**Na**: Kohl and Kohl].

Lines that are straight on a sphere (or other surfaces) are often called *geodesics*. This leads us to consider the concept of intrinsic or geodesic curvature versus extrinsic curvature. As an outside observer looking at the sphere in 3-space, all paths on the sphere, even the great circles, are curved — that is, they exhibit *extrinsic* curvature. But relative to the surface of the sphere (*intrinsically*), the lines may be straight. Be sure to understand this difference and to see why all symmetries (such as reflections) must be carried out intrinsically, or from the bug's point of view.

It is natural for you to have some difficulty experiencing straightness on surfaces other than the 2-dimensional plane; it is likely that you will start to look at spheres and the curves on spheres as 3-dimensional objects. Imagining that you are a 2-dimensional bug walking on a sphere helps you to shed your limiting extrinsic 3-dimensional vision of the curves on a sphere and to experience straightness intrinsically. Ask yourself

- ♦ What does the bug have to do, when walking on a non-planar surface, in order to walk in a straight line?
- ♦ How can the bug check if it is going straight?

Experimentation with models plays an important role here. Working with models that *you create* helps you to experience great circles as, in fact, the only straight lines on the surface of a sphere. Convincing yourself of this notion will involve recognizing that straightness on the plane and straightness on a sphere have common elements. When you are comfortable with "great-circle-straightness," you will be ready to transfer the symmetries of straight lines on the plane to great circles on a sphere and, later, to geodesics on other surfaces. Here are some activities that you can try, or visualize, to help you experience great circles and their intrinsic straightness on a sphere. However, it is better for you to come up with your own experiences.

♦ Stretch something elastic on a sphere. It will stay in place on a great circle, but it will not stay on a small circle if the sphere is slippery. Here, the elastic follows a path that is approximately the shortest because a stretched elastic always moves so that it will be shorter. This a very useful practical criterion of straightness.

♦ Roll a ball on a straight chalk line (or straight on a freshly painted floor!). The chalk (or paint) will mark the line of contact on the sphere, and it will form a great circle.

♦ Take a stiff ribbon or strip of paper that does not stretch, and lay it "flat" on a sphere. It will only lie properly along a great circle. Do you see how this property is related to local symmetry? This is sometimes called the *Ribbon Test*. (For further discussion of the Ribbon Test, see Problems **3.4** and **7.6** of [**DG**: Henderson].)

♦ The feeling of turning and "non-turning" comes up. Why is it that on a great circle there is no turning and on a latitude line there is turning? Physically, in order to avoid turning, the bug has to move its left feet the same distance as its right feet. On a non-great circle (for example, a latitude line that is not the equator), the bug has to walk faster with the legs that are on the side closer to the equator. This same idea can be experienced by taking a small toy car with its wheels fixed so that, on a plane, it rolls along a straight line. On a sphere, the car will roll around a great circle; but it will not roll around other curves.

 ◆ Also notice that, on a sphere, straight lines are circles (points
   on the surface a fixed distance away from a given point) —
   special circles whose circumferences are straight! Note that
   the equator is a circle with two intrinsic centers: the north
   pole and the south pole. In fact, any circle (such as a latitude
   circle) on a sphere has two intrinsic centers.

These activities will provide you with an opportunity to investigate
the relationships between a sphere and the geodesics of that sphere.
Along the way, your experiences should help you to discover how great
circles on a sphere have most of the same symmetries as straight lines on
a plane.

**You should pause and not read further until you have
expressed your thinking and ideas about this problem.**

SYMMETRIES OF GREAT CIRCLES

Reflection-through-itself symmetry: We can see this globally by placing a hemisphere on a flat mirror. The hemisphere together with the image in the mirror exactly recreates a whole sphere. Figure 2.2 shows a reflection through the great circle g.

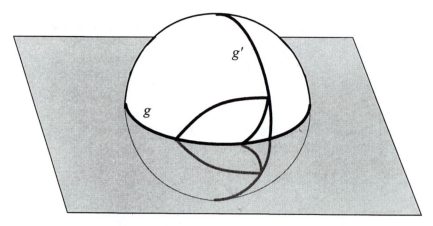

Figure 2.2   Reflection-through-itself symmetry

Reflection-perpendicular-to-itself symmetry: A reflection through any great circle will take any great circle (for example, g' in Figure 2.2) perpendicular to the original great circle onto itself.

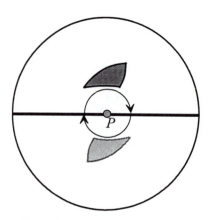

Figure 2.3   Half-turn symmetry

Half-turn symmetry: A rotation through half of a full revolution about any point P on a great circle interchanges the part of the great circle on one side of P with the part on the other side of P. See Figure 2.3.

Rigid-motion-along-itself symmetry: For great circles on a sphere, we call this a translation along the great circle or a rotation around the poles of that great circle. This property of being able to move rigidly along itself is not unique to great circles because any circle on the sphere will also have the same symmetry.

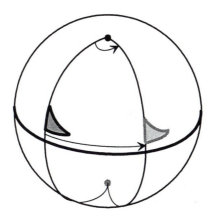

Figure 2.4   Rigid-motion-along-itself symmetry

Central symmetry, or point symmetry: Viewed intrinsically (from the 2-dimensional bug's point-of-view), central symmetry through a point P on the sphere sends any point A to the point at the same great circle distance from P but on the opposite side. See Figure 2.5.

Extrinsically (from our 3-dimensional point-of-view) central symmetry through P would send A to a point off the surface of the sphere as shown on the right in Figure 2.5. The only extrinsic central symmetry of the sphere (and the only one for great circles on the sphere) is through the center of the sphere (which is not *on* the sphere). The transformation that is intrinsically central symmetry is extrinsically half-turn symmetry (about the diameter through P). Intrinsically, as on a plane, central symmetry does not differ from half-turn symmetry with respect to the end result. This distinction between intrinsic and extrinsic is important to experience at this point.

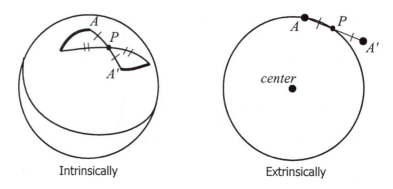

Figure 2.5   Central symmetry through P

3-dimensional-rotation symmetry: This symmetry does not hold for great circles in 3-space; however, it does hold for great circles in a 3-sphere. See Chapter 20.

You will probably notice that other objects on the sphere, besides great circles, have some of the symmetries mentioned here. It is important for you to construct such examples, and to attempt to find an object that has all of the symmetries mentioned here but is not a great circle. This will help you to realize that straightness and the five symmetries discussed here are intimately related.

*EVERY GEODESIC IS A GREAT CIRCLE

Notice that you were not asked to prove that *every geodesic on the sphere is a great circle*. This is true but more difficult to prove. Many texts simply *define* the great circles to be the straight lines (geodesics) on the sphere. We have not taken that approach. We have shown that the great circles are geodesics and it is clear that two points on the sphere are always joined by a great circle arc, which shows that there are sufficient great circle geodesics to do the geometry we wish.

To show that great circles are the only geodesics involves some notions from Differential Geometry. In Problem **3.2b** of [**DG**: Henderson] this is proved using special properties of plane curves. More generally, a geodesic satisfies a differential equation with the initial condition being a point on the geodesic and the direction of the geodesic at that point (see Problem **8.4b** of [**DG**: Henderson]). Thus it follows from the analysis theorem on the *Existence and Uniqueness of Solutions to Differential Equations* that

> **THEOREM 2.1.** *At every point and in every direction on a smooth surface there is a unique geodesic going from that point in that direction.*

From this it follows that all geodesics on a sphere are great circles. Do you see why?

*INTRINSIC CURVATURE

You have tried wrapping the sphere with a ribbon and noticed that the ribbon will only lie flat along a great circle. (If you haven't experienced this yet, then do it now before you go on.) Arcs of great circles are the only paths of a sphere's surface that are tangent to a straight line on a piece of paper wrapped around the sphere.

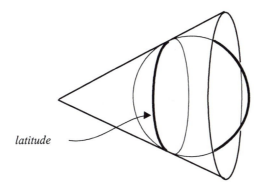

latitude

Figure 2.6   Finding the intrinsic curvature

If you wrap a piece of paper tangent to the sphere around a latitude circle (see Figure 2.6), then, extrinsically, the paper will form a portion of a cone and the curve on the paper will be an arc of a circle when the paper is flattened. The *intrinsic curvature* of a path on the surface of a sphere can be defined as the curvature that one gets when one "unwraps" the path onto a plane. For more details see Chapter 3 of [**DG:** Henderson].

Differential geometers often talk about intrinsically straight paths (geodesics) in terms of the velocity vector of the motion as one travels at a constant speed along that path. (The velocity vector is tangent to the

curve along which the bug walks.) For example, as you walk along a
great circle, the velocity vector to the circle changes direction, extrinsically, in 3-space where the change in direction is toward the center of the
sphere. "Toward the center" is not a direction that makes sense to a
2-dimensional bug whose whole universe is the surface of the sphere.
Thus, the bug does not experience the velocity vectors as changing direction at points along the great circle; however, along non-great circles the
velocity vector will be experienced as changing in the direction of the
closest center of the circle. In differential geometry, the rate of change,
from the bug's point of view, is called the *covariant* (or *intrinsic*)
derivative. As the bug traverses a geodesic, the covariant derivative of
the velocity vector is zero. This can also be expressed in terms of *parallel transport*, which is discussed in Chapters 7, 8, and 10 of this text. For
discussion of differential geometry see [**DG**: Henderson].

Chapter 3

WHAT IS AN ANGLE?

A (plane) *angle* is the inclination to one another of two lines in
a plane which meet on another and do not lie in a straight line.
— Euclid, *Elements*, Definition 8 [Appendix A]

In this chapter you will be thinking about angles. It is not necessary to do
the problems in order — you may find it easier to do Problem **3.2** before
3.1. It may help to think about the properties of angles before trying to
prove theorems about them. In a sense, you should be working on **3.1**
and **3.2** at the same time because they are so closely intertwined. This
provides a valuable opportunity to apply and reflect on what you have
learned about straightness in Chapters 1 and 2. This will also be helpful
in the further study of straightness in Chapters 4 and 5, but, if you wish,
you may study this chapter after Chapters 4 and 5.

PROBLEM **3.1** VERTICAL ANGLE THEOREM (VAT)

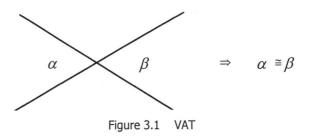

Figure 3.1   VAT

*Prove: A pair of opposite angles formed by two intersecting
straight lines are congruent.* [Note: Angles such as α and β are
called **vertical angles**.] *What properties of straight lines and/or*

25

the plane are you using in your proof? Does your proof also work on a sphere? Why?

Show how you would "move" α to make it coincide with β. We do not have in mind a formal two-column proof as used in high school geometry. As discussed in the Message to the Reader, mathematicians in actual practice usually use "proof" to mean "a convincing communication that answers — Why?"

PROBLEM 3.2   WHAT IS AN ANGLE?

Give some possible definitions of the term "angle." Do all of these definitions apply to the plane as well as to spheres? What are the advantages and disadvantages of each?

What does it mean for two angles to be congruent? How can we check?

SUGGESTIONS

Textbooks usually give some variant of the definition: *An angle is the union of two rays (or segments) with a common endpoint.* If we start with two straight line segments with a common endpoint and then add squiggly parts onto the ends of each one, would we say that the angle has changed as a result? Likewise, look at the angle formed at the lower-left-hand corner of this piece of paper. Even first grade students will recognize this as an example of an angle. Now, tear off the corner (at least in your imagination). Is the angle still there, on the piece you tore off? Now tear away more of the sides of the angles, being careful not to tear through the corner. The angle is still there at the corner, isn't it?

Figure 3.2   Where is the angle?

So what part of the angle determines how large the angle is, or if it is an angle at all? What is the angle? It seems it cannot be merely a union of two rays. Here is one of the many cases where children seem to know more than we do. Paying attention to these insights, can we get better definitions of angle? Do not expect to find one formal definition that is completely satisfactory; it seems likely that no formal definition can capture all aspects of our experience of what an angle is.

Symmetries were an important element of your solutions for Problems **1.1** and **2.1**. They will be very useful for this problem, as well. It is perfectly valid to think about measuring angles in this problem, but proofs utilizing line symmetries are generally simpler. It often helps to think of the vertical angles as whole geometric figures. Also, keep in mind that there are many different ways of looking at angles, so there are many ways of proving the vertical angle theorem. Make sure that your notions of angle and angle congruency in Problem **3.2** are consistent with your proofs in Problem **3.1**, and vice versa.

There are at least three different perspectives from which one can define "angle," as follows:

- a *dynamic* notion of angle — angle as movement;

- angle as *measure*; and,

- angle as *geometric shape*.

Each of these, separately or together, might help you prove the Vertical Angle Theorem. A *dynamic* notion of angle involves an action: a rotation, a turning point, or a change in direction between two lines. Angle as *measure* may be thought of as the arc length of a circular sector or the ratio between areas of circular sectors. Thought of as a *geometric shape,* an angle may be seen as the delineation of space by two intersecting lines. Each of these perspectives carries with it methods for checking angle congruency. You can check the congruency of two dynamic angles by verifying that the actions involved in creating or replicating them are the same. If you feel that an angle is a measure, then you must verify that both angles have the same measure. If you describe angles as geometric shapes, then one angle should be made to coincide with the other using isometries in order to prove angle congruence. Which of the above definitions has the most meaning for you? Are there any other useful ways of describing angles?

Note that we sometimes talk about ***directed angles***, or angles with direction. When considered as directed angles, we say that the angles α and β in Figure 3.3 are not the same but have equal magnitude and opposite directions (or sense). Note the similarity to the relationship between line segments and vectors.

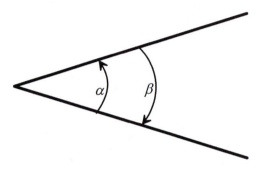

Figure 3.3   Directed angles

You should pause and not read further until you have expressed your own thinking and ideas about Problems 3.1 and 3.2.

HINTS FOR THREE DIFFERENT PROOFS

In the following section, we will give hints for three different proofs of the Vertical Angle Theorem. Note that a particular notion of angle is assumed in each proof. Pick one of the proofs, or find your own different proof that is consistent with a notion of angle and angle congruence that is most meaningful to you.

1st proof:

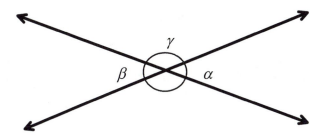

Figure 3.4   VAT using angle as measure

Each line creates a 180° angle. Thus, $\alpha + \gamma = \beta + \gamma$.

Therefore we can conclude that $\alpha \cong \beta$. But why is this so? Is it always true that if one subtracts a given angle from two 180° angles then the remaining angles are congruent?

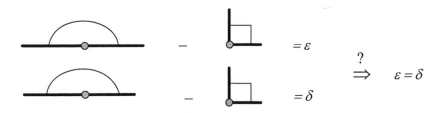

Figure 3.5   Subtracting angles and measures

Numerically, it does not make any difference how one subtracts an angle, but geometrically it makes a big difference. Behold Figure 3.6! Here, ε really cannot be considered the same as δ. Thus, measure does not completely express what we *see* in the geometry of this situation. If you wish to salvage this notion of angle as measure, then you must explain *why* it is that in this proof of the Vertical Angle Theorem γ can be subtracted from both sides of the equation $\alpha + \gamma = \beta + \gamma$.

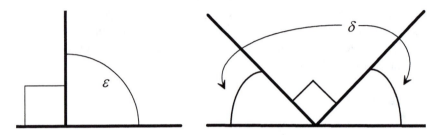

Figure 3.6 δ is not the same as ε

2nd proof: Consider two overlapping lines and choose any point on them. Rotate one of the lines, maintaining the point of intersection and making sure that the other line remains fixed.

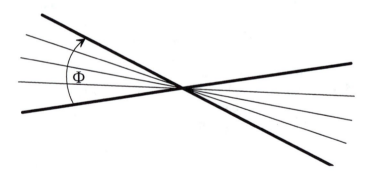

Figure 3.7   VAT using angle as rotation

What happens? What notion of angle and angle congruency is at work here?

3rd proof: What symmetries will take α onto β? See Figure 3.1 or 3.4. Use the properties of straight lines you investigated in Chapters 1 and 2.

PROBLEM **3.3** DUALITY BETWEEN POINTS AND LINES

Consider the following pairs of dual statements:

* A *straight line* has many *points* lying on it.

  A *point* has many *straight lines* lying on (intersecting) it.

These are dual statements in the sense that, if we replace "point" by "straight line" and vice versa, then the statement is still true if we also change "points lying on" to "straight lines intersecting". We say that "points" and "straight lines" are **dual** and that "point lying on a straight line" is dual to "straight line intersecting a point".

* Two points on a *straight line* determine a *line segment*.

  Two lines on a *point* determine an *angle*.

Here we have the additional dual pair of "line segment" and "angle".

* The bisector of a *line segment* is a point such that reflection through this point is a symmetry of the segment.

  The bisector of an *angle* is a straight line such that reflection through this line is a symmetry of the angle.

Here we see that "bisector of a line segment" and "bisector of an angle" are dual.

Note that the notion of duality is similar to analogy in literature.

a. *Explain how the two statements in the second bullet have a similar ambiguity on spheres.*

b. *Explain the ways in which these statements are dual*:

  Translations are related to straight lines.
  Rotations are related to points.

c. *Show that this pair of statements is incorrect as stated and then expand them to make them correct on plane and spheres.*

> *Two points determine a unique straight line.*
> *Two straight lines determine a unique point.*

d. *Find other pairs of statements that express duality between straight lines and points. (For example, think of triangles.)*

The duality expressed here between points and lines is in some ways cleaner on a sphere. We will explore related dualities in Chapters 10, 19, and 21.

Chapter 4

STRAIGHTNESS ON CYLINDERS AND CONES

> **Definition 10:** When a straight line intersects another straight line such that the adjacent angles are equal to one another, then the equal angles are called *right angles* and the lines are called *perpendicular straight lines*.
> **Postulate 4:** All right angles are equal.
> — Euclid, *Elements*, [Appendix A]

When I was in high school geometry class I could not understand why Euclid would have made such a postulate — How could they possibly *not* be equal? In this chapter you will discover that sometimes right angles are not all equal and that this is connected to cones and cylinders.

We continue with straightness, but now the goal is to think intrinsically. You should be comfortable with straightness as a *local intrinsic notion* — this is the bug's view. This notion of straightness is also the basis for the notion of *geodesics* in differential geometry. Chapters 4 and 5 can be covered in either order, but we think that the experience with cylinders and cones in **4.1** will help the reader to understand the hyperbolic plane in **5.1**. If the reader is comfortable with straightness as a local intrinsic notion, then it is also possible to skip Chapter 4 if Chapters 17 and 22 on geometric manifolds are not going to be covered. However, we suggest that the reader read the sections of this chapter starting with *Geodesics on Cylinders* (at least enough to find out what Euclid's Fourth Postulate has to do with cones and cylinders).

When looking at great circles on the surface of a sphere, we were able (except in the case of central symmetry) to see all the symmetries of straight lines from global extrinsic points of view. For example, a great circle extrinsically divides a sphere into two hemispheres that are mirror images of each other. Thus on a sphere, it is a natural tendency to use the more familiar and comfortable extrinsic lens instead of taking the bug's local and intrinsic point of view. However, on a cone and cylinder you must use the local, intrinsic point of view because there is no extrinsic view that will work except in special cases.

PROBLEM 4.1   STRAIGHTNESS ON CYLINDERS AND CONES

a. *What lines are straight with respect to the surface of a cylinder or a cone? Why? Why not?*

b. *Examine:*

- *Can geodesics intersect themselves on cylinders and cones?*

- *Can there be more than one geodesic joining two points on cylinders and cones?*

- *What happens on cones with varying cone angles, including cone angles greater than 360°?*

SUGGESTIONS

Problem **4.1** is similar to Problem **2.1**, but this time the surfaces are cylinders and cones.

Make paper models, but consider the cone or cylinder as continuing indefinitely with no top or bottom (except, of course, at the cone point). Again, imagine yourself as a bug whose whole universe is a cone or cylinder. As the bug crawls around on one of these surfaces, what will the bug experience as straight? As before, paths that are straight with respect to a surface are often called the "geodesics" for the surface.

As you begin to explore these questions, it is likely that many other related geometric ideas will arise. Do not let seemingly irrelevant excess geometric baggage worry you. Often, you will find yourself getting lost in a tangential idea, and that's understandable. Ultimately, however, the exploration of related ideas will give you a richer understanding of the scope and depth of the problem. In order to work through possible confusion on this problem, try some of the following suggestions others have found helpful. Each suggestion involves constructing or using models of cones and cylinders.

- ◆ You may find it helpful to explore cylinders first before beginning to explore cones. This problem has many aspects, but focusing at first on the cylinder will simplify some things.

- ◆ If we make a cone or cylinder by rolling up a sheet of paper, will "straight" stay the same for the bug when we unroll it? Conversely, if we have a straight line drawn on a sheet of paper and roll it up, will it continue to be experienced as straight for the bug crawling on the paper?

- ◆ Lay a stiff ribbon or straight strip of paper on a cylinder or cone. Convince yourself that it will follow a straight line with respect to the surface. Also, convince yourself that straight lines on the cylinder or cone, when looked at locally and intrinsically, have the same symmetries as on the plane.

- ◆ If you intersect a cylinder by a flat plane and unroll it, what kind of curve do you get? Is it ever straight? (One way to see this curve is to dip a paper cylinder into water.)

- ◆ On a cylinder or cone, can a geodesic ever intersect itself? How many times? This question is explored in more detail in Problem **16.1**, which the interested reader may turn to now.

- ◆ Can there be more than one geodesic joining two points on a cylinder or cone? How many? Is there always at least one? Again this question is explored in more detail in Problem **16.1**.

There are several important things to keep in mind while working on this problem. First, **you absolutely must make models**. If you attempt to visualize lines on a cone or cylinder, you are bound to make claims that you would easily see are mistaken if you investigated them on an actual cone or cylinder. Many students find it helpful to make models using transparencies.

Second, as with the sphere, you must think about lines and triangles on the cone and cylinder in an intrinsic way — always looking at things from a bug's point of view. We are not interested in what's happening in

3-space, only what you would see and experience if you were restricted to the surface of a cone or cylinder.

And last, but certainly not least, you must look at cones of different shapes, that is, cones with varying cone angles.

CONES WITH VARYING CONE ANGLES

Geodesics behave differently on differently shaped cones. So an important variable is the cone angle. The **cone angle** is generally defined as the angle measured around the point of the cone on the surface. Notice that this is an intrinsic description of angle. The bug could measure a cone angle (in radians) by determining the circumference of an intrinsic circle with center at the cone point and then dividing that circumference by the radius of the circle. We can determine the cone angle extrinsically in the following way: Cut the cone along a **generator** (a line on the cone through the cone point) and flatten the cone. The measure of the cone angle is then the angle measure of the flattened planar sector.

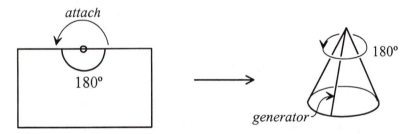

Figure 4.1   Making a 180° cone

For example, if we take a piece of paper and bend it so that half of one side meets up with the other half of the same side, we will have a 180-degree cone. A 90° cone is also easy to make — just use the corner of a paper sheet and bring one side around to meet the adjacent side.

Also be sure to look at larger cones. One convenient way to do this is to make a cone with a variable cone angle. This can be accomplished by taking a sheet of paper and cutting (or tearing) a slit from one edge to the center. (See Figure 4.2.) A rectangular sheet will work but a circular sheet is easier to picture. Note that it is not necessary that the slit be straight!

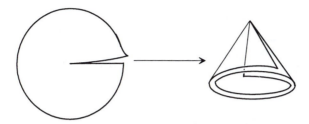

Figure 4.2   A cone with variable cone angle (0 – 360°)

You have already looked at a 360° cone — it's just a plane. The cone angle can also be larger than 360°. A common larger cone is the 450° cone. You probably have a cone like this somewhere on the walls, floor, and ceiling of your room. You can easily make one by cutting a slit in a piece of paper and inserting a 90° slice (360° + 90° = 450°).

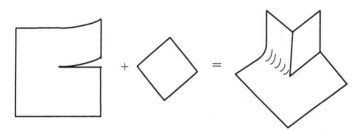

Figure 4.3   How to make a 450° cone

You may have trouble believing that this is a cone, but remember that just because it cannot hold ice cream does not mean it is not a cone. If the folds and creases bother you, they can be taken out — the cone will look ruffled instead. It is important to realize that when you change the shape of the cone like this (that is, by ruffling), you are only changing its extrinsic appearance. Intrinsically (from the bug's point of view) there is no difference. You can even ruffle the cone so that it will hold ice cream if you like, although changing the extrinsic shape in this way is not useful to a study of its intrinsic behavior.

It may be helpful for you to discuss some definitions of a cone, such as the following: *Take any simple (non-intersecting) closed curve **a** on a sphere and the center **P** of the sphere. A **cone** is the union of the rays that start at **P** and go through each point on **a**.* The cone angle is then equal to (length of **a**)/(radius of sphere), in radians. Do you see why?

You can also make a cone with variable angle of more than 180°: Take two sheets of paper and slit them together to their centers as in Figure 4.4. Tape the right side of the top slit to the left side of the bottom slit as pictured. Now slide the other sides of the slits. Try it!

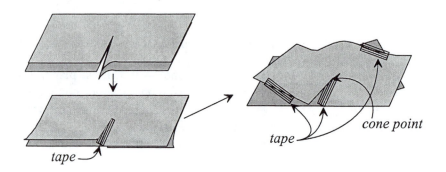

Figure 4.4   Variable cone angle larger than 360°

Experiment by making paper examples of cones like those shown above. What happens to the triangles and lines on a 450° cone? Is the shortest path always straight? Does every pair of points determine a straight line?

Finally, also consider line symmetries on the cone and cylinder. Check to see if the symmetries you found on the plane will work on these surfaces, and remember to think intrinsically and locally. A special class of geodesics on the cone and cylinder are the generators. These are the straight lines that go through the cone point on the cone or go parallel to the axis of the cylinder. These lines have some extrinsic symmetries (*can you see which ones?*), but in general, geodesics have only local, intrinsic symmetries. Also, on the cone, think about the region near the cone point — what is happening there that makes it different from the rest of the cone?

It is best if you experiment with paper models to find out what geodesics look like on the cone and cylinder before reading further.

GEODESICS ON CYLINDERS

Let us first look at the three classes of straight lines on a cylinder. When walking on the surface of a cylinder, a bug might walk along a vertical generator.

Figure 4.5   Vertical generators are straight

It might walk along an intersection of a horizontal plane with the cylinder, what we will call a *great circle*.

Figure 4.6   Great circles are intrinsically straight

Or, the bug might walk along a spiral or helix of constant slope around the cylinder.

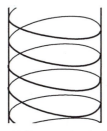

Figure 4.7   Helixes are intrinsically straight

Why are these geodesics? How can you convince yourself? And why are these the only geodesics?

GEODESICS ON CONES

Now let us look at the classes of straight lines on a cone.

Walking along a generator: When looking at straight paths on a cone, you will be forced to consider straightness at the cone point. You might decide that there is no way the bug can go straight once it reaches the cone point, and thus a straight path leading up to the cone point ends there. Or you might decide that the bug can find a continuing path that has at least some of the symmetries of a straight line. Do you see which path this is? Or you might decide that the straight continuing path(s?) is the limit of geodesics that just miss the cone point.

Help! Where do I go now?

Figure 4.8   Bug walking straight over the cone point

Walking straight and around: If you use a ribbon on a 90° cone, then you can see that this cone has a geodesic like the one depicted in Figure 4.9. This particular geodesic intersects itself. However, check to see that this property depends on the cone angle. In particular, if the cone angle is more than 180°, then geodesics do not intersect themselves. And if the cone angle is less than 90°, then geodesics (except for generators) intersect at least two times. Try it out! Later, in Chapter 17, we will describe a tool that will help you determine how the number of self-intersections depends on the cone angle.

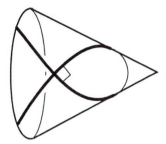

Figure 4.9   A geodesic intersecting itself on a 90° cone

LOCALLY ISOMETRIC

By now you should realize that when a piece of paper is rolled or bent into a cylinder or cone, the bug's local and intrinsic experience of the surface does not change except at the cone point. Extrinsically, the piece of paper and the cone are different, but in terms of the local geometry intrinsic to the surface they differ only at the cone point.

Two geometric spaces, **G** and **H**, are said to be ***locally isometric*** at points G in **G** and H in **H** if the local intrinsic experience at G is the same as the experience at H. That is, there are neighborhoods of G and H that are identical in terms of intrinsic geometric properties. A cylinder and the plane are locally isometric (at every point) and the plane and a cone are locally isometric except at the cone point. Two cones are locally isometric at their cone points only if the cone angles are the same.

Because cones and cylinders are locally isometric with the plane, locally they have the same geometric properties. We look at this more in Chapter 17. Later, we will show that a sphere is not locally isometric with the plane — *be on the lookout for a result that will imply this*.

IS "SHORTEST" ALWAYS "STRAIGHT"?

We are often told that "a straight line is the shortest distance between two points," but, is this really true?

As we have already seen on a sphere, two points not opposite each other are connected by two straight paths (one going one way around a great circle and one going the other way). Only one of these paths is shortest. The other is also straight, but not the shortest straight path.

Consider a model of a cone with angle 450°. Notice that such cones appear commonly in buildings as so-called "outside corners" (see Figure 4.10). It is best, however, to have a paper model that can be flattened.

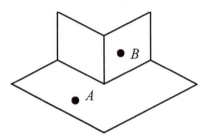

Figure 4.10   There is no straight (symmetric) path from *A* to *B*

Use your model to investigate which points on the cone can be joined by straight lines (in the sense of having reflection-in-the-line symmetry). In particular, look at points such as those labeled *A* and *B* in Figure 4.10. Convince yourself that there is no path from *A* to *B* that is straight (in the sense of having reflection-in-the-line symmetry), and for these points the shortest path goes through the cone point and thus is not straight (in the sense of having symmetry).

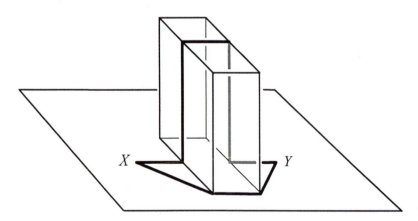

Figure 4.11   The shortest path is not straight (in the sense of symmetry)

Here is another example: Think of a bug crawling on a plane with a tall box sitting on that plane (refer to Figure 4.11). This combination surface — the plane with the box sticking out of it — has eight cone points. The four at the top of the box have 270° cone angles, and the four at the bottom of the box have 450° cone angles (180° on the box and 270° on the plane). What is the shortest path between points X and Y, points on opposite sides of the box? Is the straight path the shortest? Is the shortest path straight? To check that the shortest path is not straight, try to see that at the bottom corners of the box the two sides of the path have different angular measures. (In particular, if X and Y are close to the box, then the angle on the box side of the path measures a little more than 180° and the angle on the other side measures almost 270°.)

RELATIONS TO DIFFERENTIAL GEOMETRY

So, we see that sometimes a straight path is not shortest and the shortest path is not straight. Does it then makes sense to say (as most books do) that in Euclidean geometry a straight line is the shortest distance between two points? In differential geometry, on "smooth" surfaces, "straight" and "shortest" are more nearly the same. A *smooth* surface is essentially what it sounds like. More precisely, a surface is smooth at a point if, when you zoom in on the point, the surface becomes indistinguishable from a flat plane. (For details of this definition, see Problem 4.1 in [**DG**: Henderson].) Note that a cone is not smooth at the cone point, but a sphere and a cylinder are both smooth at every point. The following is a theorem from differential geometry:

> **THEOREM 4.1:** *If a surface is smooth then an intrinsically straight line (geodesic) on the surface is always the shortest path between "nearby" points. If the surface is also complete (every geodesic on it can be extended indefinitely), then any two points can be joined by a geodesic that is the shortest path between them.* See [**DG**: Henderson], Problem **7.4b** and **7.4d**.

Consider a planar surface with a hole removed. Check that for points near opposite sides of the hole, the shortest path (on the plane surface with hole removed) is not straight because the shortest path must go around the hole.

We encourage the reader to discuss how each of the previous examples and problems is in harmony with this theorem.

Note that the statement "every geodesic on it can be extended indefinitely" is a reasonable interpretation of Euclid's Second Postulate:

Every limited straight line can be extended indefinitely to a (unique) straight line. [Appendix A]

In addition, Euclid defines a right angle as follows:

*When a straight line intersects another straight line such that the adjacent angles are equal to one another, then the equal angles are called **right angles**.* [Appendix A]

Note that if you use this definition, then right angles at a cone point are not equal to right angles at points where the cone is locally isometric to the plane. And Euclid goes on to state as his Fourth Postulate:

All right angles are equal.

Thus, Euclid's Fourth Postulate rules out cones and any surface with isolated cone points. What is further ruled out by Euclid's Fourth Postulate would depend on formulating more precisely just what it says. It is not clear (at least to the author!) whether there is something we would want to call a surface that could be said to satisfy Euclid's Fourth Postulate and not be a smooth surface. However, it is clear that Euclid's postulate at least gives part of the meaning of "smooth surface," because it rules out isolated cone points.

Chapter 5

STRAIGHTNESS ON HYPERBOLIC PLANES

[To son János:] For God's sake, please give it [work on
hyperbolic geometry] up. Fear it no less than the sensual
passion, because it, too, may take up all your time and deprive
you of your health, peace of mind and happiness in life.

— Wolfgang Bolyai (1775–1856)
[**SE**: Davis and Hersch, page 220]

We now study hyperbolic geometry. This chapter may be skipped if the
reader will not be covering geometric manifolds and the shape of space
in Chapters 17 and 22 and if in the remainder of this book the reader
leaves out all mentions of hyperbolic planes. However, to skip studying
hyperbolic planes would be to skip an important notion in the history of
geometry, and to skip the geometry which may be the basis of our physical universe.

As with the cone and cylinder, we must use an intrinsic point of
view on hyperbolic planes. This is especially true because there is no
standard extrinsic embedding of a hyperbolic plane into 3-space.

A SHORT HISTORY OF HYPERBOLIC GEOMETRY

Hyperbolic geometry, discovered more than 170 years ago by C.F. Gauss
(1777–1855, German), János Bolyai (1802–1860, Hungarian), and N.I.
Lobatchevsky (1792–1856, Russian), is special from a formal axiomatic
point of view because it satisfies all the postulates (axioms) of Euclidean
geometry except for the parallel postulate. In hyperbolic geometry
straight lines can converge toward each other without intersecting

45

(violating Euclid's Fifth Postulate), and there is more than one straight line through a given point that does not intersect (is parallel to) a given line (violating Playfair's Parallel Postulate). (See Figure 5.1.)

The reader can explore more details of the axiomatic nature of hyperbolic geometry in Chapter 10. Note that the 450° cone also violates the two parallel postulates mentioned above. Thus the 450° cone has many of the properties of the hyperbolic plane.

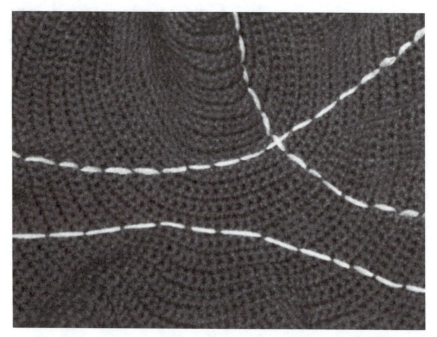

Figure 5.1   Two geodesics through a point not intersecting a given line

Hyperbolic geometry has turned out to be useful in various branches of higher mathematics. Also, the geometry of binocular visual space appears experimentally to be best represented by hyperbolic geometry (see [**NE**: Zage]). In addition, hyperbolic geometry is one of the possible geometries for our three-dimensional physical universe — we will explore this connection more in Chapters 17 and 22.

Hyperbolic geometry and non-Euclidean geometry are considered in many books as being synonymous, but as we have seen there are other non-Euclidean geometries, particularly spherical geometry. It is also not accurate to say (as many books do) that non-Euclidean geometry was discovered about 170 years ago. Spherical geometry (which is clearly not Euclidean) was in existence and studied by at least the ancient Babylonians, Indians, and Greeks more than 2,000 years ago. Spherical geometry was of importance for astronomical observations and astrological calculations. In Aristotle we can find evidence that non-Euclidean geometry was studied even before Euclid. (See [**Hi**: Heath, page 57] and [**Hi**: Toth].) Even Euclid in his *Phaenomena* [**AT**: Euclid] (a work on astronomy) discusses propositions of spherical geometry. Menelaus, a Greek of the first century, published a book *Sphaerica*, which contains many theorems about spherical triangles and compares them to triangles on the Euclidean plane. (*Sphaerica* survives only in an Arabic version. For a discussion see [**Hi**: Kline, page 119–120].)

Most texts and popular books introduce hyperbolic geometry either axiomatically or via "models" of the hyperbolic geometry in the Euclidean plane. These models are like our familiar map projections of the earth and (like these maps of the earth) intrinsic straight lines on the hyperbolic plane (surface of the earth) are not, in general, straight in the model (map) and the model, in general, distorts distances and angles. We will return to the subject of projection and models in Chapter 16.

In this chapter we will introduce the geometry of the hyperbolic plane as the intrinsic geometry of a particular surface in 3-space, in much the same way that we introduced spherical geometry by looking at the intrinsic geometry of the sphere in 3-space. Such a surface is called an *isometric embedding* of the hyperbolic plane into 3-space. We will construct such a surface in the next section. Nevertheless, many texts and popular books say that David Hilbert (1862–1943, German) proved in 1901 that it is not possible to have an isometric embedding of the hyperbolic plane onto a closed subset of Euclidean 3-space. These authors miss what Hilbert actually proved. In fact, Hilbert [**NE**: Hilbert] proved that there is no *real analytic* isometry (that is, no isometry defined by real-valued functions which have convergent power series). In 1972, Tilla Milnor [**NE**: Milnor] extended Hilbert's result by proving that there is no isometric embedding defined by functions whose first and second derivatives are continuous. Without giving an explicit construction, N. Kuiper [**NE**: Kuiper] showed in 1955 that there is a differentiable isometric embedding onto a closed subset of 3-space.

The construction used here was shown to the author by William Thurston (b.1946, American) in 1978[†]; and it is not defined by equations

---

[†]The idea for this construction is also included in Thurston's recent book [**DG**: Thurston, pages 49 and 50] and is discussed in the author's recent book [**DG**: Henderson, page 31].

at all, because it has no definite embedding in Euclidean space. In Problem **5.2** we will show that our isometric model is locally isometric to a certain smooth surface of revolution called the ***pseudosphere***, which is well known to locally have hyperbolic geometry. Later, in Chapter 16, we will explore the various (non-isometric) models of the hyperbolic plane (these models are the way that hyperbolic geometry is presented in most texts) and prove that these models and the isometric constructions here produce the same geometry.

CONSTRUCTIONS OF HYPERBOLIC PLANES

We will describe four different isometric constructions of hyperbolic planes (or approximations to hyperbolic planes) as surfaces in 3-space. It is very important that you actually perform at least one of these constructions. The act of constructing the surface will give you a feel for hyperbolic planes that is difficult to get any other way. Templates for all the paper constructions (and information about possible availability of crocheted hyperbolic planes) can be found at the supplements site

www.math.cornell.edu/~dwh/books/eg00/supplements.html

1. THE HYPERBOLIC PLANE FROM PAPER ANNULI

A paper model of the hyperbolic plane may be constructed as follows: Cut out many identical annular ("annulus" is the region between two concentric circles) strips as in Figure 5.2. Attach the strips together by taping the inner circle of one to the outer circle of the other. It is crucial that all the annular strips have the same inner radius and the same outer radius, but the lengths of the annular strips do not matter. You can also cut an annular strip shorter or extend an annular strip by taping two strips together along their straight ends. The resulting surface is of course only an approximation of the desired surface. The actual hyperbolic plane is obtained by letting $\delta \to 0$ while holding the radius ρ fixed. Note that since the surface is constructed (as $\delta \to 0$) the same everywhere it is ***homogeneous*** (that is, intrinsically and geometrically, every point has a neighborhood that is isometric to a neighborhood of any other point). We will call the results of this construction the ***annular hyperbolic plane.*** I strongly suggest that the reader take the time to **cut out carefully several such annuli and tape them together as indicated.**

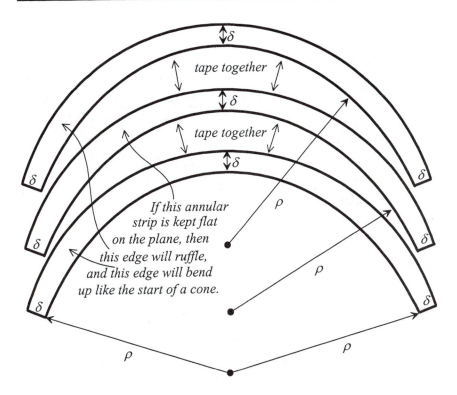

Figure 5.2   Annular strips for making an annular hyperbolic plane

2. How to Crochet the Hyperbolic Plane

Once you have tried to make your annular hyperbolic plane from paper annuli you will certainly realize that it will take a lot of time. Also, later you will have to play with it carefully because it is fragile and tears and creases easily — you may want just to have it sitting on your desk. But there is another way to get a sturdier model of the hyperbolic plane, which you can work and play with as much as you wish. This is the crocheted hyperbolic plane.

In order to make the crocheted hyperbolic plane you need just very basic crocheting skills. All you need to know is how to make a chain (to start) and how to single crochet. That's it! Now you can start. See Figure 5.3 for a picture of these stitches, and see their description in the list below.

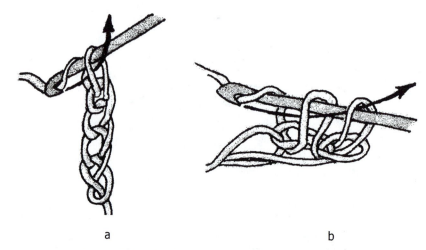

a   b

Figure 5.3   Crochet stitches for the hyperbolic plane

First you should choose a yarn that will not stretch a lot. Every yarn will stretch a little but you need one that will keep its shape. Now you are ready to start the stitches:

1. Make your **beginning chain stitches** (Figure 5.3a). About 20 chain stitches for the beginning will be enough.

2. **For the first stitch in each row** insert the hook into the 2nd chain from the hook. Take yarn over and pull through chain, leaving 2 loops on hook. Take yarn over and pull through both loops. One single crochet stitch has been completed. (Figure 5.3b.)

3. **For the next *N* stitches** proceed exactly like the first stitch except insert the hook into the next chain (instead of the 2nd).

4. **For the (*N* + 1)st stitch** proceed as before except insert the hook into the same loop as the *N*-th stitch.

5. **Repeat Steps 3 and 4** until you reach the end of the row.

6. **At the end of the row** before going to the next row do one extra chain stitch.

7. **When you have the model as big as you want,** you can stop by just pulling the yarn through the last loop.

Be sure to crochet fairly tightly and evenly. That's all you need from crochet basics. Now you can go ahead and make your own hyperbolic plane. You have to increase (by the above procedure) the number of stitches from one row to the next in a constant ratio, N to N + 1 — the ratio and size of the yarn determine the radius (the ρ in the annular hyperbolic plane) of the hyperbolic plane. You can experiment with different ratios BUT not in the same model. We suggest that you start with a ratio of 5 to 6. You will get a hyperbolic plane ONLY if you will be increasing the number of stitches in the same ratio all the time.

Crocheting will take some time but later you can work with this model without worrying about destroying it. The completed product is pictured in Figure 5.4.

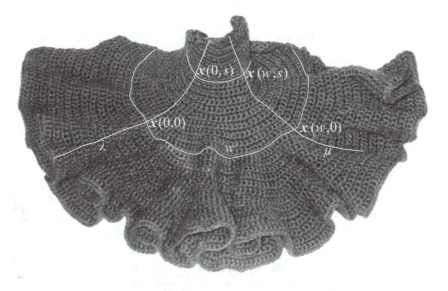

Figure 5.4   A crocheted annular hyperbolic plane

3. {3,7} AND {7,3} POLYHEDRAL CONSTRUCTIONS

A polyhedral model can be constructed from equilateral triangles by putting 7 triangles together at every vertex, or by putting 3 regular heptagons (7-gons) together at every vertex. These are called the {3,7} **polyhedral model** and the {7,3} **polyhedral model** because triangles (3-gons) are put together 7 at a vertex, or heptagons (7-gons) are put together 3 at a vertex. These models have the advantage of being constructed more easily than the annular or crocheted models; however, one cannot make better and better approximations by decreasing the size of

the triangles. This is true because at each vertex the cone angle is (7 ×
$\pi/3$) = 420° or (3 × $5\pi/7$) = 385.71...°), no matter what the size of the
triangles and heptagons are; whereas the hyperbolic plane in the small
looks like the Euclidean plane with 360° cone angles. Another disadvantage of the polyhedral model is that it is not easy to describe the annuli
and related coordinates.

You can make these models less "pointy" by replacing the sides of
the triangles with arcs of circles in such a way that the new vertex angles
are $2\pi/7$, or by replacing the sides of the heptagons with arcs of circles
in such a way that the new vertex angles are $2\pi/3$. But then the model is
less easy to construct because you are cutting and taping along curved
edges.

See Problems **10.6** and **21.5** for more discussions of regular polyhedral tilings of plane, spheres, and hyperbolic planes.

4. HYPERBOLIC SOCCER BALL CONSTRUCTION

We now explore a polyhedral construction that involves two different
regular polygons instead of the single polygon used in the {3,7} and
{7,3} polyhedral constructions. A spherical soccer ball (outside the
USA, called a football) is constructed by using pentagons surrounded by
five hexagons or two hexagons and one pentagon together around each
vertex. The plane can be tiled by hexagons, each surrounded by six other
hexagons. The hyperbolic plane can be approximately constructed by
using heptagons (7-sided) surrounded by seven hexagons and two
hexagons and one heptagon together around each vertex. See Figure 5.5.
Because a heptagon has interior angles with $5\pi/7$ radians (= 128.57...°),
the vertices of this construction have cone angles of 368.57...° and thus
are much smoother than the {3,7} and {7,3} polyhedral constructions. It
also has a nice appearance if you make the heptagons a different color
from the hexagons. It is also easy to construct (as long as you have a
template — you can find a variety on the supplements website). As with
any polyhedral construction one cannot get closer and closer approximations to the hyperbolic plane. There is also no apparent way to see the
annuli.

The hyperbolic soccer ball construction is related to the {3,7} construction in the sense that if a neighborhood of each vertex in the {3,7} construction is replaced by a heptagon then the remaining portion of each triangle is a hexagon.

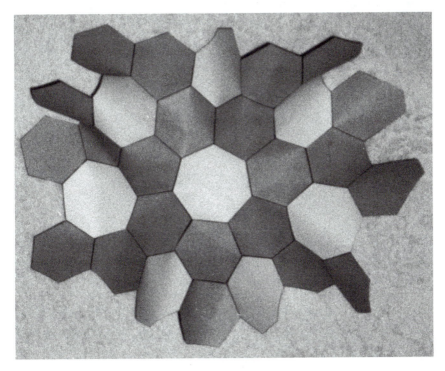

Figure 5.5   The hyperbolic soccer ball

5. "{3,6½}" Polyhedral Construction

We can avoid some of the disadvantages of the {3,7} and soccer ball constructions by constructing a polyhedral annulus. In this construction we have seven triangles together only at every other vertex and six triangles together at the others. This construction still has the disadvantage of not being able to produce closer and closer approximations and it also is more "pointy" (larger cone angles) than the hyperbolic soccer ball.

The precise construction can be described in two different (but, in the end, equivalent) ways:

1. Construct polyhedral annuli as in Figure 5.6 and then tape them together as with the annular hyperbolic plane.

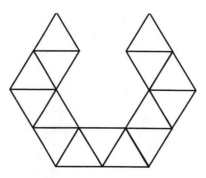

Figure 5.6   Polyhedral annulus

2. The quickest way is to start with many strips as pictured in Figure 5.7a — these strips can be as long as you wish. Then add four of the strips together as in Figure 5.7b, using five additional triangles. Next, add another strip every place there is a vertex with five triangles and a gap (as at the marked vertices in Figure 5.7b). Every time a strip is added an additional vertex with seven triangles is formed.

The center of each strip runs perpendicular to each annulus, and you can show that these curves (the center lines of the strip) are each geodesics because they have local reflection symmetry.

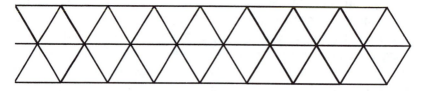

Figure 5.7a   Strips

Figure 5.7b   Forming the polyhedral annular hyperbolic plane

HYPERBOLIC PLANES OF DIFFERENT RADII (CURVATURE)

Note that the construction of a hyperbolic plane is dependent on ρ (the radius of the annuli), which is often called the *radius of the hyperbolic plane*. As in the case of spheres, we get different hyperbolic planes depending on the value of ρ. In Figures 5.8–5.10 (see next page) there are crocheted hyperbolic planes with radii approximately 4 cm, 8 cm, and 16 cm. The pictures were all taken from approximately the same perspective and in each picture there is a centimeter rule to indicate the scale.

Note that as ρ increases the hyperbolic plane becomes flatter and flatter (has less and less curvature). For both the sphere and the hyperbolic plane as ρ goes to infinity they both become indistinguishable from the ordinary flat (Euclidean) plane. Thus, the plane can be called a sphere (or hyperbolic plane) with infinite radius. In Chapter 7, we will define the "Gaussian Curvature" and show that it is equal to $1/\rho^2$ for a sphere and $-1/\rho^2$ for a hyperbolic plane.

Figure 5.8   Hyperbolic plane with $\rho \approx 4$ cm

Figure 5.9   Hyperbolic plane with $\rho \approx 8$ cm

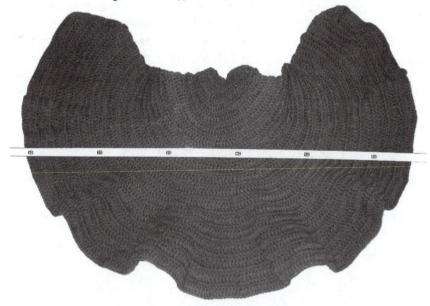

Figure 5.10   Hyperbolic plane with $\rho \approx 16$ cm

PROBLEM **5.1**   WHAT IS STRAIGHT IN A HYPERBOLIC PLANE?

a. *On a hyperbolic plane, consider the curves that run radially across each annular strip. Argue that these curves are intrinsically straight. Also, show that any two of them are asymptotic, in the sense that they converge toward each other but do not intersect.*

Look for the local intrinsic symmetries of each annular strip and then global symmetries in the whole hyperbolic plane. Make sure you give a convincing argument why the symmetry holds in the limit as $\delta \to 0$.

We shall say that two geodesics that converge in this way are **asymptotic geodesics**. Note that there are no geodesics (straight lines) on the plane that are asymptotic.

b. *Find other geodesics on your physical hyperbolic surface. Use the properties of straightness (such as symmetries) you talked about in Problems **1.1**, **2.1**, and **4.1**.*

Try holding two points between the index fingers and thumbs on your two hands. Now pull gently — a geodesic segment with its reflection symmetry should appear between the two points. If your surface is durable enough, try folding the surface along a geodesic. Also, you may use a ribbon to test for geodesics.

c. *What properties do you notice for geodesics on a hyperbolic plane? How are they the same as geodesics on the plane or spheres, and how are they different from geodesics on the plane and spheres?*

Explore properties of geodesics involving intersecting, uniqueness, and symmetries. Convince yourself as much as possible using your model — full proofs for some of the properties will have to wait until Chapter 16.

*PROBLEM **5.2**   THE PSEUDOSPHERE IS HYPERBOLIC

Show that locally the annular hyperbolic plane is isometric to portions of a (smooth) surface defined by revolving the graph of

*a continuously differentiable function of z about the z-axis. This
is the surface usually called the **pseudosphere**.*

OUTLINE OF PROOF

1. Argue that each point on the annular hyperbolic plane is like any
   other point. (Think of the annular construction.)

2. Start with one of the annular strips and complete it to a full annulus
   in a plane. Then, construct a surface of revolution by attaching to
   the inside edge of this annulus other annular strips as described in
   the construction of the annular hyperbolic plane. (See Figure 5.11.)
   Note that the second and subsequent annuli form truncated cones.
   Finally, imagine the width of the annular strips, δ, shrinking to zero.

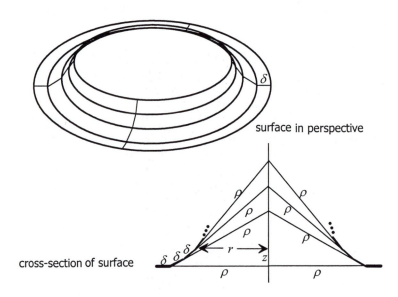

Figure 5.11   Hyperbolic surface of revolution — pseudosphere

3. Derive a differential equation representing the coordinates of a
   point on the surface using the geometry inherent in Figure 5.11. If
   $f(r)$ is the height (z-coordinate) of the surface at a distance of r from
   the z-axis, then the differential equation should be (remember that ρ
   is a constant)

$$\frac{dz}{dr} = \frac{-r}{\sqrt{\rho^2 - r^2}}$$

4. Solve (using tables or computer algebra systems) the differential equation for z as a function of r. Note that you are not getting r as a function of z.

5. Then argue (using a theorem from first-semester calculus) that r is a continuously differentiable function of z.

We can also crochet a pseudosphere by starting with 5 or 6 chain stitches and continuing in a spiral fashion, increasing as when crocheting the hyperbolic plane. See Figure 5.12. Note that, when you crochet beyond the annular strip that lays flat and forms a complete annulus, the surface forms ruffles and is no longer a surface of revolution (nor a smooth surface).

Figure 5.12   Crocheted pseudosphere

The term "pseudosphere" seems to have originated with Hermann von Helmholtz (1821–1894, German) who was contrasting spherical space with what he called pseudospherical space. However, Helmholtz did not actually find a surface with this geometry. Eugenio Beltrami (1835–1900, Italian) actually constructed the surface, which is called the pseudosphere, and showed that its geometry is locally the same as (locally isometric to) the hyperbolic geometry constructed by Lobatchevsky. (For more historical discussion, see [**Hi**: Katz], pages 781–83.) Mathematicians searched further for a surface (in those days "surface" meant "real analytic surface") that would be the whole of the hyperbolic plane (as opposed to only being locally isometric to it). This search was halted when Hilbert proved that such a surface was impossible in his theorem that we discussed above at the end of the section *A Short History of Hyperbolic Geometry* on page 47.

PROBLEM 5.3   ROTATIONS AND REFLECTIONS ON SURFACES

On the plane or on spheres, rotations and reflections are both *intrinsic* in the sense that they are experienced by a 2-dimensional bug as rotations and reflections. These intrinsic rotations and reflections are also *extrinsic* in the sense that they can also be viewed as isometries of 3-space. (For example, the reflection of a sphere through a great circle can also be viewed as a reflection of 3-space through the plane containing the great circle.) Thus rotations and reflections are particularly easy to see on planes and spheres. In addition, on the plane and sphere all rotations and reflections are *global* in the sense that they take the whole plane to itself or whole sphere to itself. (For example, any intrinsic rotation about a point on a sphere is always a rotation of the whole sphere.) On cylinders and cones, intrinsic rotations and reflections exist locally because cones and cylinders are locally isometric with the plane. However

    a. *Which intrinsic rotations and reflections on which cones and cylinders are also extrinsic? Which are global?*

Be sure to look at the cone points. The answers to the two questions are not exactly the same.

Now, we can see from our physical hyperbolic planes that geodesics exist joining every pair of points and that these geodesics each have reflection-in-themselves symmetry. (If you did not see this in Problem **5.1c**, then go back and explore some more with your physical model. In Chapter 16 we will prove rigorously that this is in fact true by using the upper half plane model.) In Chapter 16 we will show that these reflections are global reflections of the whole hyperbolic space. However, note that there do not exist extrinsic reflections of the hyperbolic plane (in Euclidean 3-space). However, given all this, it is not clear that there exist intrinsic rotations, nor is it necessarily clear what exactly intrinsic rotations are.

b. *Let l and m be two geodesics on the hyperbolic plane which intersect at the point P. Look at the composition of the reflection R_l through l with the reflection R_m through m. Show that this composition R_mR_l deserves to be called a rotation about P. What is the angle of the rotation?*

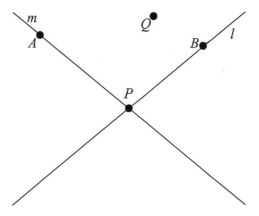

Figure 5.13   Composition of two reflections is a rotation

Let *A* be a point on *m* and *B* be a point on *l*, and let *Q* be an arbitrary point (not on *m* or *l*). Investigate where *A*, *B*, and *Q* are sent by R_l and then by R_mR_l . See Figure 5.13.

We will study symmetries and isometries in more detail in Chapter 11. In that chapter we will show that every isometry (on the plane, spheres, and hyperbolic planes) is a composition of one, two, or three reflections.

    c. *Show that Problem* **3.1** *(VAT) holds on cylinders, cones (including the cone points), and hyperbolic planes.*

If you check your proof(s) of **3.1** and modify them (if necessary) to only involve symmetries, then you will be able to see that they hold also on the other surfaces.

    d. *Define "rotation about P through an angle θ" without mentioning reflections in your definition.*

What does a rotation do to a point not at *P*?

    e. *A popular high school textbook series **defines** a rotation as the composition of two reflections. Is this a good definition? Why? or Why not?*

Chapter 6

TRIANGLES AND CONGRUENCIES

> *Polygons* are those figures whose boundaries are made
> of straight lines: *triangles* being those contained by three, ...
>
> Things which coincide with one another are equal to
> one another.
>
> — Euclid, *Elements*, Definition 19 &
> Common Notion 4 [Appendix A]

At this point, you should be thinking intrinsically about the surfaces of spheres, cylinders, cones, and hyperbolic planes. In the problems to come you will have opportunities to apply your intrinsic thinking when you make your own definitions for triangle on the different surfaces and investigate congruence properties of triangles.

In this chapter we will begin our study of triangles and their congruencies on all the surfaces that you have studied: plane, spheres, cones, cylinders, and hyperbolic spaces. [If you skipped any of these surfaces, you should find that this and the succeeding chapters will still make sense, but you will want to limit your investigations to triangles on the surfaces you studied.]

Before starting with triangles we must first discuss a little more general information about geodesics.

*GEODESICS ARE LOCALLY UNIQUE

In previous chapters we have studied geodesics, intrinsically straight paths. Our main criterion has been that a path is intrinsically straight (and thus, a geodesic) if it has local intrinsic reflection-through-itself symmetry. Using this notion we found that joining any pair of points there is a geodesic that, on a sphere, is a great circle and, on a hyperbolic space, has reflection-through-itself symmetry. However, on more general surfaces, which may have no (even local) reflections, it is necessary to have a deeper definition of geodesic in terms of intrinsic curvature. (See for example, Chapter 3 of [**DG**: Henderson].) Then, to be precise, we must prove that the geodesics we found on spheres and hyperbolic planes are the only geodesics on these surfaces. It is easy to see that these geodesics that we have found are enough to have, for every point and every direction from that point, one geodesic proceeding from that point in that direction. To prove that these are the *only* geodesics it is necessary (as we have mentioned before) to involve some notions from Differential Geometry. In particular, one must first define a notion of geodesic that will work on general surfaces which have no (even local) intrinsic reflections. Then one shows that a geodesic satisfies a second order (nonlinear) differential equation (see Problem **8.4b** of [**DG**: Henderson]). Thus, it follows from the analysis theorem on the *Existence and Uniqueness of Differential Equations*, with the initial conditions being a point on the geodesic and the direction of the geodesic at that point, that

>   **THEOREM 6.0.** *For any given point and any direction at that point on a smooth surface there is a unique geodesic starting at that point and going in the given direction.*

From this it follows that the geodesics with local intrinsic reflection-in-itself symmetry, which we found in Problems **2.1**, **4.1**, and **5.1**, are all the geodesics on spheres, cylinders, cones, and hyperbolic planes.

PROBLEM 6.1   PROPERTIES OF GEODESICS

In this problem we ask you to pull together a summary of the properties of geodesics on the plane, spheres, and hyperbolic planes. Mostly, you have already argued that these are true but we summarize the results here to remind us what we have seen and so that you can reflect again about why these are true. Remember that cylinders and cones (not at the cone point) are locally the same geometrically as (locally isometric to) the plane; thus, geodesics on the cone and cylinder are locally (but not globally) the same as straight lines on the plane. We will return to a study of geodesics on cylinders and cones in Chapter 17.

a.   *For every geodesic on the plane, sphere, and hyperbolic plane there is a reflection of the whole space through the geodesic.*

b.   *Every geodesic on the plane, sphere, and hyperbolic plane can be extended indefinitely* (in the sense that the bug can walk straight ahead indefinitely along any geodesic).

c.   *Every pair of distinct points on the plane, sphere, and hyperbolic plane determines a* (not necessarily unique) *geodesic.*

d.   *Every pair of distinct points on the plane or hyperbolic plane determines a unique geodesic segment joining them. On the sphere there are always at least two such segments.*

e.   *On the plane or on a hyperbolic plane, two geodesics either coincide or are disjoint or they intersect in one point. On a sphere, two geodesics either coincide or intersect exactly twice.*

Note that for the plane and hyperbolic plane, Part **d** and Part **e** are equivalent in the sense that they each imply the other.

Notice that these properties distinguish a sphere from both the Euclidean plane and from a hyperbolic plane; however, these properties do not distinguish the plane from a hyperbolic plane.

PROBLEM **6.2**   ISOSCELES TRIANGLE THEOREM (ITT)

In order to start out with some common ground, let us agree on some terminology: A *triangle* is a geometric figure formed of three points (*vertices*) joined by three straight line (geodesic) segments (*sides*). A triangle divides the surface into two regions (the *interior* and *exterior*). The (*interior*) *angles* of the triangle are the angles between the sides in the interior of the triangle. (As we will discuss below, on a sphere you must decide which region you are going to call the interior — often the choice is arbitrary.)

We will find the Isosceles Triangle Theorem very useful in studying circles and the other congruence properties of triangles because the two congruent sides can be considered to be radii of a circle.

a. (*ITT*) *Given a triangle with two of its sides congruent, then are the two angles opposite those sides also congruent? Look at this on all five of the surfaces we are studying.*

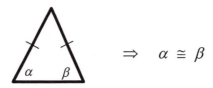

Figure 6.1   ITT

Use symmetries to solve this problem. Also, look for counterexamples — if there were counterexamples, what could they look like? If you think that ITT is not true for all triangles on a particular surface, then describe a counterexample **and** look for a smaller class of triangles that do satisfy ITT. In the process of these investigations you will need to use properties of straight lines (geodesics) on the various surfaces (see Problem **6.1**). State explicitly what properties you are using.

In your proof of Part **a**, try to see that you have also proved the following:

b. COROLLARY. *The bisector of the top angle of an isosceles triangle is also the perpendicular bisector of the base of that triangle.*

You may also want to prove a converse of ITT, but we will not use it in this book:

*c. CONVERSE OF **ITT**. *If two angles of a triangle are congruent, then are the sides opposite these angles also congruent?*

Use symmetry and look out for counterexamples — they do exist for the converse.

CIRCLES

To study congruencies of triangles we will need to know something about circles and constructions of bisectors and perpendicular bisectors.

We define a **circle** intrinsically.

A circle is a geometric figure formed by rotating one endpoint of a geodesic segment about its other endpoint, which stays fixed.

The intrinsic radius of the circle is the rotating segment and the intrinsic center of the circle is the endpoint about which the rotation is fixed. Note that on a sphere every circle has two (intrinsic) centers which are antipodal (and, in general, two different radii).

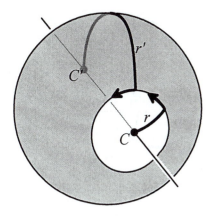

Figure 6.2   Circles on a sphere have two centers

Now ITT can be used to prove theorems about circles. For example,

THEOREM 6.2. *On the plane, spheres, hyperbolic planes, and locally on cylinders and cones, if the centers of two circles are disjoint (and not antipodal), then the circles intersect in either 0, 1, or 2 points. If the centers of the two circles coincide (or are antipodal), then the circles either coincide or are disjoint.*

Proof: Because cylinders and cones are locally isometric to the plane, locally and intrinsically circles will behave the same as on the plane. Thus we limit the remainder of this proof to the plane, spheres, and hyperbolic planes. Let C and C' denote the centers of the circles. See Figure 6.3.

If C and C' are antipodal on a sphere and the two circles intersect at P, then a (extrinsic) plane through P (perpendicular to the extrinsic diameter CC') will intersect the sphere in a circle that must coincide with the two given circles. If C and C' coincide on a sphere and the circles intersect, then pick the antipodal point to C as the center of the first circle, which reduces this to the case we just considered. If C and C' coincide on the plane and hyperbolic planes and the circles intersect, then the circles have the same radii because two points are joined by only one line segment. Thus, if the centers coincide or are antipodal, the circles coincide or are disjoint.

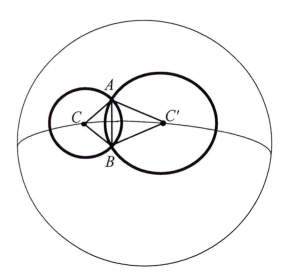

Figure 6.3   Intersection of two circles

Thus, we can now assume that C and C' are disjoint and not antipodal, so there is a unique geodesic joining the centers. If A and B are two points of intersection of the circles, then $\triangle ACB$ and $\triangle AC'B$ are isosceles triangles.

But, given that $\triangle ACB$ and $\triangle AC'B$ are isosceles, the Corollary to ITT asserts that the bisectors of $\angle ACB$ and $\angle AC'B$ must be perpendicular bisectors of their common base. Thus, the union of the two angle bisectors is straight and joins C and C'. So, the union must be contained in the unique geodesic determined by C and C'. Therefore, any pair of intersections of the two circles, such as A and B, must lie on opposite sides of this unique great circle. Immediately, it follows that there cannot be more than two intersections.

PROBLEM **6.3**   BISECTOR CONSTRUCTIONS

a. *Show how to use a compass and straight edge to construct the perpendicular bisector of a straight line segment. How do you know it is actually the perpendicular bisector? How does it work on the sphere?*

Use ITT and Theorem **6.2**. Be sure that you have considered all segment lengths on the sphere.

b. *Show how to use a compass and straight edge to find the bisector of any angle. How do you know it actually is the angle bisector? How does it work on the sphere?*

Use ITT and Part **a**. Be sure that you have considered all sizes of angles.

It is a part of mathematical folklore that it is impossible to trisect an angle with compass and straight edge, but we will show in Problem **11.4** that, in fact, it is possible. In addition we will discuss what facts the folklore refers to.

PROBLEM **6.4**   SIDE-ANGLE-SIDE (**SAS**)

Are two triangles congruent if two sides and the included angle of one are congruent to two sides and the included angle of the other?

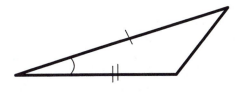

Figure 6.4   SAS

In some textbooks SAS is listed as an axiom; in others it is listed as the definition of congruency of triangles, and in others as a theorem to be proved. But no matter how one considers SAS, it still makes sense and is important to ask: Why is SAS true on the plane?

Is SAS true on spheres, cylinders, cones, and hyperbolic planes?

If you find that SAS is not true for all triangles on a sphere or another surface, is it true for sufficiently small triangles? Come up with a definition for "small triangles" for which SAS does hold.

SUGGESTIONS

Be as precise as possible, but **use your intuition**. In trying to prove SAS on a sphere you will realize that SAS does not hold unless some restrictions are made on the triangles. Keep in mind that everyone sees things differently, so there are many possible definitions of "small." Some may be more restrictive than others (that is, they don't allow as many triangles as other definitions). Use whatever definition makes sense for you.

Remember that it is not enough to simply state what a small triangle is; you must also prove that SAS is true for the small triangles under your definition — explain why the counterexamples you found before are now ruled out and explain why the condition(s) you list is (are) sufficient to prove SAS. Also, try to come up with a basic, general proof that can be applied to all surfaces.

And remember what we said before: By "proof" we mean what most mathematicians use in their everyday practice, i.e., a convincing communication that answers — Why? We do not ask for the usual two-column proofs from high school (unless, of course, you find the two-column proof is sufficiently convincing and answers — Why?). Your proof should convey the meaning you are experiencing in the situation. Think about why SAS is true on the plane — think about what it means for actual physical triangles — then try to translate these ideas to the other surfaces.

Let us clarify some terminology that we have found to be helpful for discussing SAS and other theorems. Two triangles are said to be *congruent* if, through a combination of translations, rotations, and reflections, one of them can be made to coincide with the other. In fact (as we will prove in Chapter 11), we only need to use reflections. If an even number of reflections are needed, then the triangles are said to be *directly congruent*. In this course we will focus on *congruence* and not specifically on *direct congruence*; however, some readers may wish to keep track of the distinction as we go along.

B'
C'
A'
B
C
A
B''
C''
A''

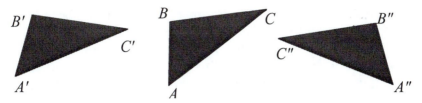

Figure 6.5 Direct congruence and congruence

In Figure 6.5, ΔABC is directly congruent to $\Delta A'B'C'$ but ΔABC is not directly congruent to $\Delta A''B''C''$. However, ΔABC is congruent to both $\Delta A'B'C'$ and $\Delta A''B''C''$ and we write: $\Delta ABC \cong \Delta A'B'C' \cong \Delta A''B''C''$.

So, why is SAS true on the plane? We will now illustrate one way of looking at this question. Referring to Figure 6.6, suppose that ΔABC and $\Delta A'B'C'$ are two triangles such that $\angle BAC \cong \angle B'A'C'$, $AB \cong A'B'$ and $AC \cong A'C'$.

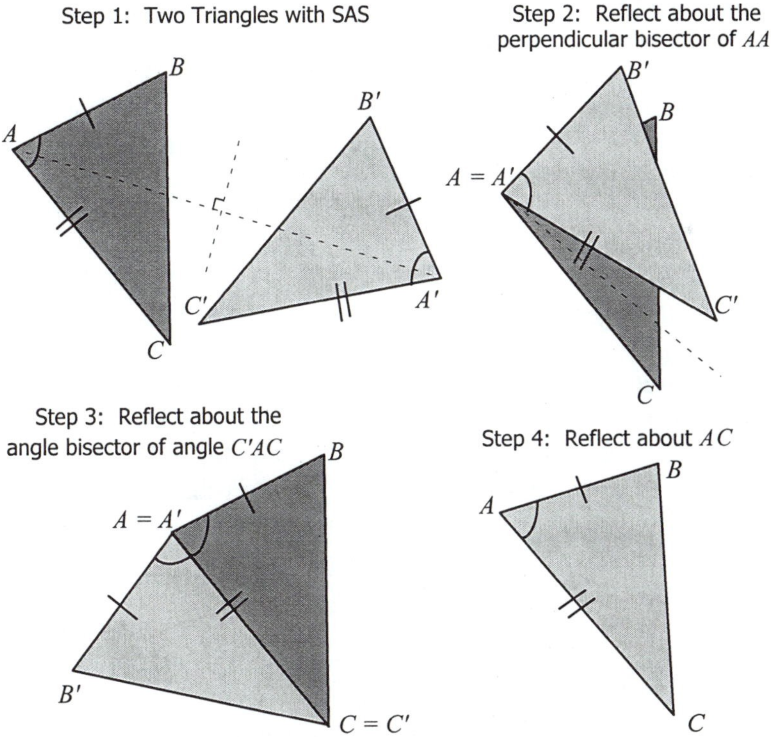

Step 1: Two Triangles with SAS
A
B
C
B'
C'
A'
Step 2: Reflect about the perpendicular bisector of AA'
B'
B
A = A'
C'
C
Step 3: Reflect about the angle bisector of angle C'AC
B
A = A'
B'
C = C'
Step 4: Reflect about AC
B
A
C

Figure 6.6 SAS on plane

Reflect $\triangle A'B'C'$ about the perpendicular bisector (Problem **6.3**) of AA' so that A' coincides with A. Because the sides AC and $A'C'$ are congruent, we can now reflect about the angle bisector of $\angle C'AC$. Now C' coincides with C. (*Why?*) If after this reflection B and B' are not coincident, then a reflection (about $AC = A'C'$) will complete the process and all three vertices, the two given sides, and the included angle of the two triangles will coincide. *So, why is it that, on the plane, the third sides (BC and $B'C'$) must now be the same?*

Because the third sides (BC and $B'C'$) coincide, $\triangle ABC$ is congruent to $\triangle A'B'C'$. (In the case that only two reflections are needed, the two triangles are directly congruent.)

The proof of SAS on the plane is not directly applicable to the other surfaces because properties of geodesics differ on the various surfaces. In particular, the number of geodesics joining two points varies from surface to surface and is also relative to the location of the points on the surface. On a sphere, for example, there are always at least two straight paths joining any two points. As we saw in Chapter 4, the number of geodesics joining two points on a cylinder is infinite. On a cone the number of geodesics is dependent on the cone angle, but for cones with angles less than 180° there is more than one geodesic joining two points. It follows that the argument made for SAS on the plane is not valid on cylinders, cones, or spheres. The question then arises: Is SAS *ever* true on those surfaces?

Look for triangles for which SAS is not true. Some of the properties that you found for geodesics on spheres, cones, cylinders, and hyperbolic planes will come into play. As you look closely at the features of triangles on those surfaces, you may find that they challenge your notions of triangle. Your intuitive notion of triangle may go beyond what can be put into a traditional definition of triangle. When you look for a definition of **small triangle** for which SAS will hold on these surfaces, you should try to stay close to your intuitive notion. In the process of exploring different triangles you may come up with examples of triangles that seem very strange. Let us look at some unusual triangles.

For instance, keep the examples in Figure 6.7 in mind. All the lines shown in Figure 6.7 are geodesic segments of the sphere. The two sides and their included angle for SAS are marked. As you can see, there are two possible geodesics that can be drawn for the third side — the short one in front and the long one that goes around the back of the sphere.

Remember that on a sphere, any two points define at least two geodesics (an infinite number if the points are at opposite poles).

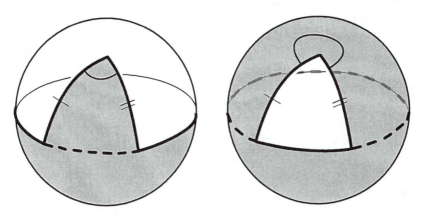

Figure 6.7   Two counterexamples for SAS on sphere

Look for similar examples on a cone and cylinder. You may decide to accept the smaller triangle into your definition of "small triangle" but to exclude the large triangle from your definition. But what is a large triangle? To answer this, let us go back to the plane. What is a triangle on the plane? What do we choose as a triangle on the plane?

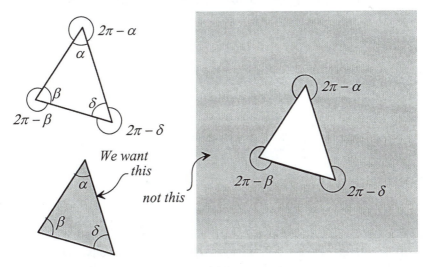

Figure 6.8   We choose the interior of a plane triangle to have finite area

On the plane, a figure that we want to call a triangle has all of its angles on the "inside." Also, there is a clear choice for *inside* on the plane; it is the side that has finite area. See Figure 6.8. But what is the inside of a triangle on a sphere?

The restriction that the area on the inside has to be finite does not work for the spherical triangles because all areas on a sphere are finite. So what is it about the large triangle that challenges our view of triangle? You might try to resolve the triangle definition problem by specifying that each side must be the shortest geodesic between the endpoints. However, be aware that antipodal points (that is, a pair of points that are at diametrically opposite poles) on a sphere do not have a unique shortest geodesic joining them. On a cylinder we can have a triangle for which all the sides are the shortest possible segments, yet the triangle does not have finite area. Try to find such an example. In addition, a triangle on a cone will always bound one region that has finite area, but a triangle that encircles the cone point may cause problems.

PROBLEM 6.5   ANGLE-SIDE-ANGLE (ASA)

Are two triangles congruent if one side and the adjacent angles of one are congruent to one side and the adjacent angles of another?

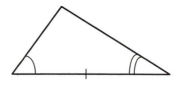

Figure 6.9   ASA

SUGGESTIONS

This problem is similar in many ways to the previous one. As before, look for counterexamples on all surfaces; and, if ASA does not hold for all triangles, see if it works for small triangles. If you find that you must restrict yourself to small triangles, see if your previous definition of "small" still works; if it does not work here, then modify it.

It is also important to keep in mind when considering ASA that both of the angles must be on the same side — the *interior* of the triangle. For example, see Figure 6.10.

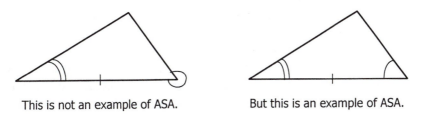

This is not an example of ASA.   But this is an example of ASA.

Figure 6.10   Angles of a triangle must be on same side

Let us look at a proof of ASA on the plane as depicted in Figure 6.11.

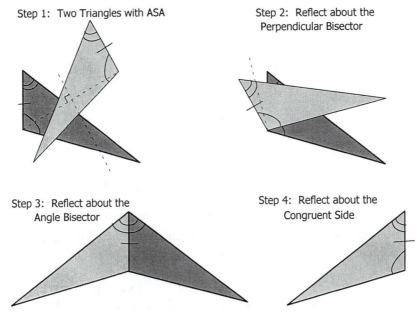

Figure 6.11   ASA on the plane

The planar proof for ASA does not work on spheres, cylinders, and cones because, in general, geodesics on these surfaces intersect in more than one point. But can you make the planar proof work on a hyperbolic plane?

As was the case for SAS, we must ask ourselves if we can find a class of small triangles on each of the different surfaces for which the above argument is valid. You should check if your previous definitions of small triangle are too weak, too strong, or just right to make ASA true on spheres, cylinders, cones, and hyperbolic planes. It is also important to look at cases for which ASA does not hold. Just as with SAS, some interesting counterexamples arise.

In particular, try out the configuration in Figure 6.12 on a sphere. To see what happens you will need to try this on an actual sphere. If you extend the two sides to great circles, what happens? You may instinctively say that it is not possible for this to be a triangle, and on the plane most people would agree, but *try it on an physical sphere and see what happens*. Does it define a unique triangle? Remember that on a sphere two geodesics always intersect twice.

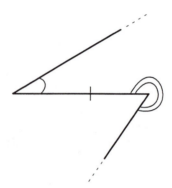

Figure 6.12   Possible counterexample to ASA

Finally, notice that in our proof of ASA on the plane, we did not use the fact that the sum of the angles in a triangle is 180°. We avoided this for two reasons. For one thing, to use this "fact" we would have to prove it first. This is both time consuming and unnecessary. We will prove it later (in different ways) in Chapters 7 and 10. More importantly, such a proof will not work on spheres and hyperbolic planes because the sum of the angles of triangles on spheres and hyperbolic planes is not always 180° — see the triangles depicted in Figures 6.13 and 6.14. We will explore further the sum of the angles of a triangle in Chapter 7.

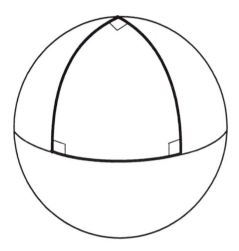

Figure 6.13   Triple-right triangle on a sphere

Figure 6.14   Hyperbolic triangle

Remember that it is best to come up with a proof that will work for all surfaces because this will be more powerful, and, in general, will tell us more about the relationship between the plane and the other surfaces.

Chapter 7

AREA AND HOLONOMY

We [my student and I] are both greatly amazed; and my share
in the satisfaction is a double one, for he sees twice over who
makes others see.
— Jean Henri Fabre, *The Life of the Fly*,
New York: Dodd, Mead and Co, 1915, p. 300.

There are many things in this chapter that have amazed me and my
students. I hope you, the reader, will also be amazed by them. We will
find a formula for the area of triangles on spheres and hyperbolic planes.
We will then investigate the connections between area and **parallel
transport**, a notion of local parallelism that is definable on all surfaces.
We will also introduce the notion of **holonomy**, which has many applications in modern differential geometry and engineering.

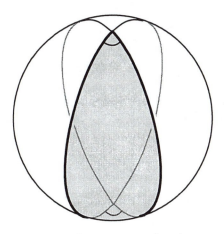

Figure 7.1   Lune or diangle

DEFINITION: A *lune* or *diangle* is any of the four regions determined by two (not coinciding) great circles (see Figure 7.1). The two angles of the lune are congruent. (*Why?*)

PROBLEM 7.1   THE AREA OF A TRIANGLE ON A SPHERE

a. *The two sides of each interior angle of a triangle Δ on a sphere determine two congruent lunes with lune angle the same as the interior angle. Show how the three pairs of lunes determined by the three interior angles, α, β, γ, cover the sphere with some overlap. (What is the overlap?)*

Draw this on a physical sphere, as in Figure 7.2.

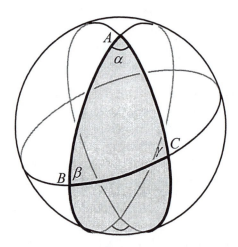

Figure 7.2   Finding the area of a spherical triangle

b. *Find a formula for the area of a lune with lune angle θ in terms of θ and the (surface) area of the sphere (of radius ρ), which you can call S_ρ. Use radian measure for angles.*

Hint: What if θ is π? π/2?

c. *Find a formula for the area of a triangle on a sphere of radius ρ.*

SUGGESTIONS

This is one of the problems that you almost certainly must do on an actual sphere. There are simply too many things to see, and the drawings we make on paper distort lines and angles too much. The best way to start is to make a small triangle on a sphere, and extend the sides of the triangle to complete great circles. Then look at what you've got. You will find an identical triangle on the other side of the sphere, and you can see several lunes that extend out from the triangles. The key to this problem is to put everything in terms of areas that you know. We will see later (Problem **14.2**) that the area of the whole sphere with radius ρ is $S_\rho = 4\pi\rho^2$, or you may find a derivation of this formula in a multivariable calculus text, or you can just leave your answer in terms of S_ρ.

PROBLEM **7.2**   AREA OF HYPERBOLIC TRIANGLES

Before we start to explore the area of a general triangle on the hyperbolic plane, we first look for triangles with large area.

a. *On your hyperbolic plane draw as large a triangle as you can find. Compare your triangle with the large triangles that others have found. What do you notice?*

This part of the problem is best to do communicating with other people.

We can try to mimic the derivation of the area of spherical triangles, but of course there are no lunes and the area of the hyperbolic plane is evidently infinite. Nevertheless, if we focus on the exterior angles of a hyperbolic triangle and look at the regions formed, we obtain a picture of the situation in the annular hyperbolic plane. See Figure 7.3. Draw this picture on your hyperbolic plane.

In Figure 7.3, a triangle is drawn with its interior angles, α, β, γ, and exterior angles, $\pi - \alpha$, $\pi - \beta$, $\pi - \gamma$. The three extra lines are geodesics that are asymptotic at both ends to an extended side of the triangle. We call the region enclosed by these three extra geodesics an ***ideal triangle***. In the annular hyperbolic plane these are not actually triangles because their vertices are at infinity. In Figure 7.3 we see that the ideal triangle is divided into the original triangle and three "triangles" that have two of their vertices at infinity. We call a "triangle" with two vertices at infinity (and all sides geodesics) a **2/3-ideal triangle**. You can use this decomposition to determine the area of a hyperbolic triangle in much the same way you determined the area of a spherical triangle. So, first we must investigate the areas of ideal and 2/3-ideal triangles.

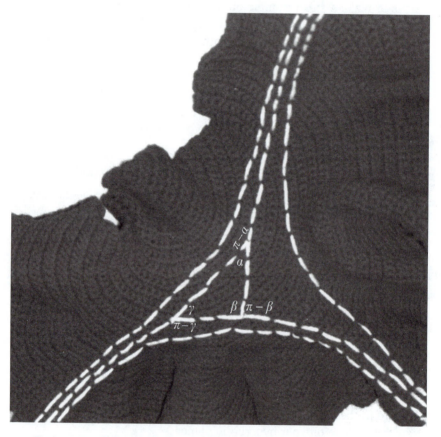

Figure 7.3   Triangle with an ideal triangle and three 2/3-ideal triangles

Now let us look at 2/3-ideal triangles.

b. *Show that on the same hyperbolic plane, all 2/3-ideal triangles with the same angle θ are congruent.*

Think of the proof of SAS (Problem **6.4**). If you have two 2/3-ideal triangles with angle θ, then by reflections you can place one of the θ-angles on top of the other. The triangles will then coincide except possibly for the third side, which is asymptotic to the two sides of the angle θ. Now you must argue that these third sides must coincide. Or, in other words, why is the situation in Figure 7.4 impossible for 2/3-ideal triangles on a hyperbolic plane?

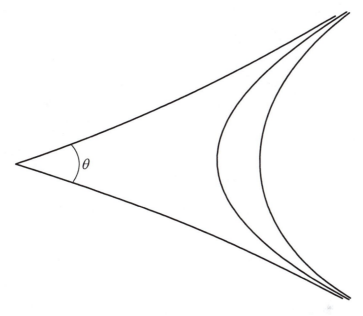

Figure 7.4. Are 2/3-ideal triangles with angle θ congruent?

Because all the 2/3-ideal triangles are congruent, we can define an area function as

$A_\rho(\alpha)$ = area of a 2/3 ideal triangle with *exterior* angle α
on a hyperbolic plane with radius ρ.

Note that $A_\rho(0) = 0$. (This is the reason we use exterior angles $\alpha = \pi - \theta$.)

c. *Show that the area function A_ρ is an additive function. That is,*

$$A_\rho(\alpha + \beta) = A_\rho(\alpha) + A_\rho(\beta).$$

Look at the picture in Figure 7.5 and show that the area of $\triangle ADE$ is the sum of the areas of triangles $\triangle ABC$ and $\triangle ACE$ by showing that $\triangle PDE$ is congruent to $\triangle PBC$.

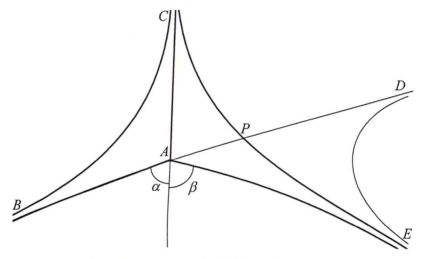

Figure 7.5   Area of 2/3-ideal triangles is additive

So we now have shown that the area function A_ρ is additive and it is also clearly continuous.

THEOREM 7.2. *A continuous additive function (from the reals to the reals) is linear.*

Because the area function is additive, it also is true that it is linear over the rationals. For example,

$$2A_\rho(\alpha) = A_\rho(\alpha) + A_\rho(\alpha) = A_\rho(\alpha + \alpha) = A_\rho(2\alpha),$$

and, if you set $\beta = 2\alpha$, then the same equations show that $\frac{1}{2}A_\rho(\beta) = A_\rho(\frac{1}{2}\alpha)$. Thus, because the area function is continuous, the function must be linear (over the reals).

Therefore, $A_\rho(\alpha) = $ constant $\times \alpha$, for $0 \le \alpha < \pi$. If we let the finite vertex of 2/3-ideal triangle go to infinity, then the interior angle will go to zero and the exterior angle will go to π. Thus $A_\rho(\pi)$ must be the area of an ideal triangle. So we have proved

All ideal triangles on the same hyperbolic plane have the same area, which we can call I_ρ.

So we can write the area function as

$$A_\rho(\alpha) = \alpha \times (I_\rho/\pi).$$

In fact, we will show in Problem **16.4** that all ideal triangles (on the same hyperbolic plane) are congruent. This is a result you may have guessed from your work in Part **a**. We will also show in Problem **16.4** that the formula for the area of an ideal triangle is $I_\rho = \pi \rho^2$. Then it follows that $A_\rho(\alpha) = \alpha \rho^2$. Notice that it is only here that we know for certain that 2/3-ideal triangles (and ideal triangles) have finite area, though you may have surmised that from your work on Part **a**.

> **d.** *Find a formula for the area of a hyperbolic triangle.*

Look at Figure 7.3 and put it together with what we have just proved.

PROBLEM **7.3**   SUM OF THE ANGLES OF A TRIANGLE

> **a.** *What can you say about the sum of the interior angles of trian-gles on spheres and hyperbolic planes? Are there maximum and/or minimum values for the sum?*

Look at triangles with non-zero area and use your formulas from Problems **7.1** and **7.2**.

> **b.** *What is the sum of the (interior) angles of a planar triangle?*

Let $\triangle ABC$ be a triangle on the plane and imagine a sphere of radius ρ passing through the points A, B, C. These three points also determine a small spherical triangle on the sphere. Now imagine the radius ρ grow-ing to infinity and the spherical triangle converging to the planar triangle.

This result for the plane is normally proved after invoking a parallel postulate. Here, we are making the assumption that the plane is a sphere of infinite radius. We will turn to a discussion of the various parallel postulates in Chapter 10.

INTRODUCING PARALLEL TRANSPORT AND HOLONOMY

Imagine that you are walking along a straight line or geodesic carrying a horizontal stick that makes a fixed angle with the line you are walking on. If you walk along the line maintaining the direction of the stick relative to the line constant, then you are performing a ***parallel transport*** of that "direction" along the path. (See Figure 7.6.)

To express the parallel transport idea, it is common terminology to say that

- *r'* is a parallel transport of *r* along *l*;

- *r* is a parallel transport of *r'* along *l*;

- *r* and *r'* are parallel transports along *l*;

- *r* can be parallel transported along *l* to *r'*; or,

- *r'* can be parallel transported along *l* to *r*.

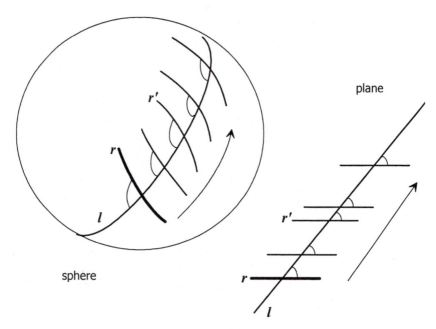

Figure 7.6   Parallel transport

On the plane there is a global notion of parallelism — two lines in the same plane are ***parallel*** if they do not intersect when extended. It follows from Problem **8.2** (or, for the plane, from standard results in high school geometry) that if two lines are parallel transports along another line in the plane or the hyperbolic plane, then they are also parallel in the sense that they will not intersect if extended. On a sphere this is not true — any two great circles on the same sphere intersect, and intersect twice. In Problem **10.1** you will show that if two lines in the plane are parallel transports along a third line, then they are parallel transports along every line that transverses them. This is also not true on a sphere and not true on a hyperbolic plane. For example, any two great circles (longitudes) through the north pole are parallel transports of each other along the equator, but they are not parallel transports along great circles near the north pole. We will explore this aspect of parallel transport more in Chapters 8 and 10.

Parallel transport has become an important notion in differential geometry, physics, and mechanics. One important aspect of differential geometry is the study of properties of spaces (surfaces) from an intrinsic point of view. As we have seen, it is not in general possible to have a global notion of direction that will determine when a direction (vector) at one point is the same as a direction (vector) at another point. However, we can say that they have the same direction *with respect to* a geodesic *g* if they are parallel transports of each other along *g*.

Parallel transport can be extended to arbitrary curves as we shall discuss at the end of this chapter. There is even a mechanical device (first developed in third-century China!), called the "South-Seeking Chariot", which will perform parallel transport along a curve on a surface. See [**DG**: Santander].

With this notion it is possible to talk about the rate at which a particular vector quantity changes intrinsically along a curve (covariant differentiation). In general, covariant differentiation is useful in the areas of physics and mechanics. In physics, the notion of parallel transport is central to some of the theories that have been put forward as possible candidates for a "Unified Field Theory," a hoped-for but as yet unrealized theory that would unify all known physical laws about forces of nature.

Let us explore what happens when we parallel transport a line segment around a triangle. For example, consider on a sphere an isosceles triangle with base on the equator and opposite vertex on the north pole (see Figure 7.7). Note that the base angles are right angles. Now start at the north pole with a vector (a directed geodesic segment — the gray arrows in Figure 7.7) and parallel transport it along one of the sides of the triangle until it reaches the base. Then parallel transport it along the base to the third side. Then parallel transport back to the north pole along the third side. Notice that the vector now points in a different direction than it did originally. You can follow a similar story for the right hyperbolic triangle represented in Figure 7.8 and see that here also there is a difference between the starting vector and the ending parallel transported vector. This difference is called the ***holonomy*** of the triangle. Note that the difference angle is counterclockwise on the sphere and clockwise in the hyperbolic plane.

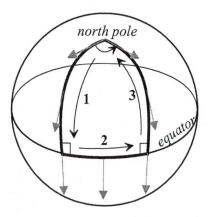

Figure 7.7   The holonomy of a double-right triangle on a sphere

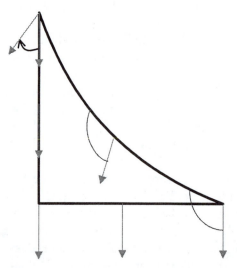

Figure 7.8   Holonomy of a hyperbolic triangle

This works for any small triangle (that is, a triangle that is contained in an open hemisphere) on a sphere and for all triangles in a hyperbolic plane. We can define *the holonomy of a* (*small*, if on a sphere) *triangle*, $\mathcal{H}(\Delta)$, as follows:

> *If you parallel transport a vector (a directed geodesic segment) counterclockwise around the three sides of a small triangle, then the holonomy of the triangle is the smallest angle from the original position of the vector to its final position with counterclockwise being positive and clockwise being negative.*

For the spherical triangle in Figure 7.7 we see that the holonomy is positive and equal to the upper angle of the triangle. For the hyperbolic triangle in Figure 7.8 we see that the holonomy is negative (clockwise).

Holonomy can also be defined for large triangles on a sphere but it is more complicated because of the confusion as to what angle to measure. For example, what should be the holonomy when you parallel transport around the equator — 0 radians or 2π radians?

PROBLEM 7.4   THE HOLONOMY OF A SMALL TRIANGLE

> *Find a formula that expresses the holonomy of a small triangle on a sphere, and a formula that expresses the holonomy of any triangle on a hyperbolic plane. What is the holonomy of a triangle on the plane?*

SUGGESTIONS

What happens to the holonomy when you change the angle at the north pole of the triangle in Figure 7.7? What happens if you parallel transport around the triangle a vector pointing in a different direction? Parallel transport vectors around different triangles on your model of a sphere. Try it on triangles that are very nearly the whole hemisphere and try it on very small triangles. What do you notice? Try this also on your models of the hyperbolic plane, again for different size triangles.

A good way to approach the formula for general triangles is to start with any geodesic segment at one of the angles of the triangle, and follow it as it is parallel transported around the triangle. Keep track of the relationships between the angles this segment makes with the sides and the exterior angles. See Figure 7.9, which is drawn for spherical triangles; the reader should be able to draw an analogous picture for a general hyperbolic triangle.

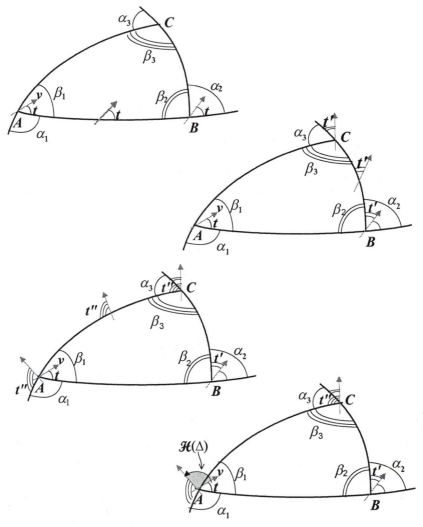

Figure 7.9   Holonomy of a general triangle

Pause, explore, and write out your ideas for this problem before reading further.

THE GAUSS–BONNET FORMULA FOR TRIANGLES

In working on Problem **7.4** you should find (among other things) that

> *The holonomy of a (small, if on a sphere) triangle is equal to 2π minus the sum of the exterior angles or equal to the sum of the interior angles minus π.*

Let $\beta_1, \beta_2, \beta_3$ be the interior angles of the triangle and $\alpha_1, \alpha_2, \alpha_3$ the exterior angles. Then algebraically the statement above can be written as:

$$\mathcal{H}(\Delta) = 2\pi - (\alpha_1 + \alpha_2 + \alpha_3) = (\beta_1 + \beta_2 + \beta_3) - \pi.$$

The quantity $[\,\Sigma\beta_i - \pi\,] = [\,2\pi - \Sigma\alpha_i\,]$ is also called the **excess** of Δ, and when the excess is negative, the positive quantity $[\,\pi - \Sigma\beta_i\,] = [\,\Sigma\alpha_i - 2\pi\,]$ is called the **defect** of Δ.

If you have not already seen it, note now the close connection between the holonomy, the excess, and the area of a triangle. Note that the holonomy is positive for triangles on a sphere and negative for triangles in a hyperbolic plane (and zero for triangles on a plane). One consequence of this formula is that the holonomy does not depend on either the vertex or the vector we start with. This is to be expected because parallel transport does not change the relative angles of any figure.

Following Problems **7.1**, **7.2**, and **7.4** we can write the result for triangles on a sphere with radius ρ in this form:

sphere:
$$\mathcal{H}(\Delta) = (\beta_1 + \beta_2 + \beta_3) - \pi = Area(\Delta)\,4\pi/S_\rho = Area(\Delta)\,\rho^{-2}.$$

For a hyperbolic plane made with annuli with radius ρ we get:

hyperbolic:
$$\mathcal{H}(\Delta) = (\beta_1 + \beta_2 + \beta_3) - \pi = -Area(\Delta)\,\pi/I_\rho = -Area(\Delta)\,\rho^{-2}.$$

The quantity ρ^{-2} is traditionally called the **Gaussian curvature** or just plain **curvature** of the sphere and $-\rho^{-2}$ is called the (**Gaussian**) **curvature** of the hyperbolic plane. If K denotes the (*Gaussian*) curvature as just defined, then the formula

$$(\beta_1 + \beta_2 + \beta_3) - \pi = Area(\Delta)\, K$$

is called the **Gauss–Bonnet Formula** (for triangles). The formula is originally due to C.F. Gauss (1777–1855, German) and was extended by P.O. Bonnet (1819–1892, French) as we will describe at the end of this chapter.

Can you see how this result gives a bug on the surface an intrinsic way of determining the quantity K and thus also determining the extrinsic radius ρ?

The Gauss–Bonnet Formula not only holds for triangles in an open hemisphere or in a hyperbolic plane but can also be extended to any simple (that is, non-intersecting) polygon (that is, a closed curve made up of a finite number of geodesic segments) contained in an open hemisphere or in a hyperbolic plane.

*PROBLEM 7.5   GAUSS–BONNET FORMULA
FOR POLYGONS

DEFINITION. *The holonomy of a simple polygon,* $\mathcal{H}(\Gamma)$, *in an open hemisphere or in a hyperbolic plane is defined as follows:*

If you parallel transport a vector (a directed geodesic segment) counterclockwise around the sides of the simple polygon, then the holonomy of the polygon is the smallest angle measured counterclockwise from the original position of the vector and its final position.

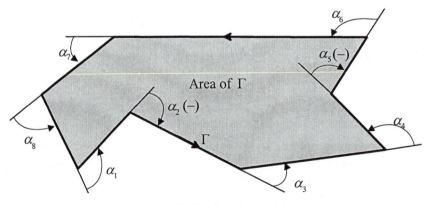

Figure 7.10   Exterior angles

If you walk around a polygon with the interior of the polygon on the left, the exterior angle at a vertex is the change in the direction at that vertex. This change is positive if you turn counterclockwise and negative if you turn clockwise. (See Figure 7.10.)

We will first look at convex polygons because this is the only case we will need later and it is easier to understand. A region is called **convex** if every pair of points in the region can be joined by a geodesic segment lying wholly in the region.

a. *Show that if Γ is a convex polygon in an open hemisphere or in a hyperbolic plane, then*

$$\mathcal{H}(\Gamma) = 2\pi - \Sigma\alpha_i = \Sigma\beta_i - (n-2)\pi = Area(\Gamma)\,K,$$

where $\Sigma\alpha_i$ is the sum of the exterior angles, $\Sigma\beta_i$ is the sum of the interior angles, n is the number of sides, and K is the Gaussian curvature.

Divide the convex polygon into triangles as in Figure 7.11. Now apply **7.4** to each triangle and carefully add up the results. You can check directly that $\mathcal{H}(\Gamma) = 2\pi - \Sigma\alpha_i$.

b. *Prove that every simple polygon on the plane or on a hemisphere or on a hyperbolic plane can be dissected into triangles without adding extra vertices.*

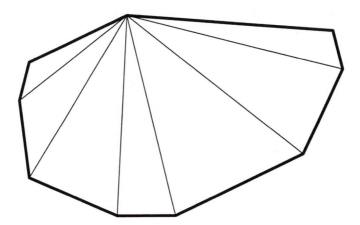

Figure 7.11   Dividing a convex polygon into triangles

SUGGESTIONS

Look at this on the plane, hemispheres, and hyperbolic planes. The difficulty in this problem is coming up with a method that works for all polygons, including very general or complex ones, such as the polygon in Figure 7.12.

Figure 7.12   General polygon

You may be tempted to try to connect nearby vertices to create triangles, but how do we know that this is always possible? How do you know that in any polygon there is even one pair of vertices that can be joined in the interior? The polygon may be so complex that parts of it get in the way of what you're trying to connect. So you might start by giving a convincing argument that there is at least one pair of vertices that can be joined by a segment in the interior of the polygon.

To see that there *is* something to prove here, Figure 7.13 shows an example of a polyhedron in 3-space with **no** pair of vertices that can be joined in the interior. The polyhedron consists of eight triangular faces and six vertices. Each vertex is joined by an edge to four of the other vertices and the straight line segment joining it to the fifth vertex lies in the exterior of the polygon. Therefore it is impossible to dissect this polyhedron into tetrahedra without adding extra vertices. This example and some history of the problem are discussed in [**Di:** Eves (1963), page 211] and [**Di:** Ho].

Note that there is at least one convex vertex (a vertex with interior angle less than π) on every polygon (in fact it is not too hard to see that there must be at least three such vertices). To see this, pick any geodesic in the exterior of the polygon and parallel transport it toward the polygon until it first touches the polygon. It is easy to see that the line must now be intersecting the polygon at a convex vertex.

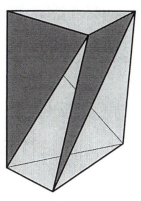

Figure 7.13   A polyhedron with vertices not joinable in the interior

c.   *Show that if* Γ *is a simple polygon in an open hemisphere or in a hyperbolic plane, then*

$$\mathcal{H}(\Gamma) = 2\pi - \Sigma\alpha_i = \Sigma\beta_i - (n-1)\pi = Area(\Gamma)\,K,$$

where $\Sigma\alpha_i$ *is the sum of the exterior angles,* $\Sigma\beta_i$ *is the sum of the interior angles, and K is the Gaussian curvature.*

Start by applying Part **b**. Then proceed as in Part **a**, but for this part you may find it easier to show that the holonomy of the polygon is the sum of the holonomies of the triangles by removing one triangle at a time. Again, you can check directly that $\mathcal{H}(\Gamma) = 2\pi - \Sigma\alpha_i$.

*GAUSS–BONNET FORMULA FOR POLYGONS ON SURFACES

The above discussion of holonomy is in the context of an open hemisphere and a hyperbolic plane, but the results have a much more general applicability and constitute an important aspect of differential geometry. In particular, we can extend this result even further to general surfaces, even those of non-constant curvature. In fact, Gauss defined the (***Gaussian***) ***curvature*** ***K(p)*** at a point *p* on any surface to be

$$K(p) = \lim_{\Delta \to p} \mathcal{H}(\Delta) \,/\, A(\Delta),$$

where the limit is taken over a sequence of small (geodesic) triangles that converge to p. The reader can check that the Gaussian curvature of a sphere (with radius ρ) is $1/\rho^2$ and that the Gaussian curvature of a hyperbolic plane (with radius ρ, the radius of the annular strips) is $-1/\rho^2$. This definition leads us to another formula, namely,

THEOREM 7.5a. *The Gauss–Bonnet Formula for Polygons on Surfaces*

On any smooth surface (2-manifold), if Γ is a (geodesic) polygon that bounds a contractible region, then

$$\mathcal{H}(\Gamma) = 2\pi - \Sigma\ \alpha_i\ = \iint_{I(\Gamma)} K(p)\ dA,$$

where the integral is the (surface) integral over $I(\Gamma)$, the interior of the polygon.

A region is said to be **contractible** if it can be continuously deformed to a point in its interior. See Figure 7.14 for examples.

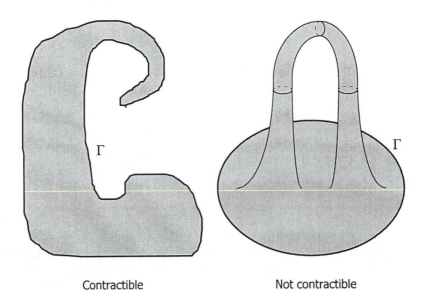

Contractible   Not contractible

Figure 7.14

The proof of this formula involves dividing the interior of Γ into many triangles, each so small that the curvature K is essentially constant over its interior, and then applying the Gauss–Bonnet Formula for spheres and hyperbolic planes to each of the triangles.

In addition, all of the versions of the Gauss–Bonnet Formula given thus far can be extended to arbitrary, simple, piece-wise smooth, closed curves. (It is this extension that was Bonnet's contribution to the Gauss–Bonnet Formula.) If γ is such a curve, then we can define the holonomy $\mathcal{H}(\gamma) = \lim \mathcal{H}(\gamma_i)$, where the limit is over a sequence (which converges point-wise to γ) of geodesic polygons $\{\gamma_i\}$ whose vertices lie on γ. Using this definition, the Gauss–Bonnet formula can be extended even further.

THEOREM 7.5b. *The Gauss–Bonnet Formula for Curves That Bound a Contractible Region*

On a sphere or hyperbolic plane, with (constant) curvature K,

$$\mathcal{H}(\gamma) = A(\gamma)\, K ,$$

where $A(\gamma)$ is the area of the region bounded by γ.

On general surfaces,

$$\mathcal{H}(\gamma) = \iint_{I(\gamma)} K(p)\, dA,$$

where $I(\gamma)$ is the interior of the region bounded by γ.

Another version of the Gauss–Bonnet Formula is discussed in Problem **17.6**, where the integral is over the whole surface.

For further discussions see the author's *Differential Geometry: A Geometric Introduction* [**DG**: Henderson], Chapters 5 and 6, especially Problems **5.4** and **6.4**.

Chapter 8

PARALLEL TRANSPORT

> *Parallel* straight lines are straight lines lying in a plane which
> do not meet if continued indefinitely in both directions.
> — Euclid, *Elements*, Definition 23 [Appendix A]

In this chapter we will further develop the notion of *parallel transport*
that was introduced in Chapter 7.

PROBLEM 8.1 EUCLID'S EXTERIOR ANGLE THEOREM (EEAT)

Any exterior angle of a triangle is greater *than each of the opposite interior angles.*[†]

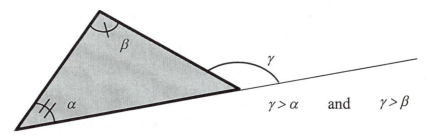

$$\gamma > \alpha \quad \text{and} \quad \gamma > \beta$$

Figure 8.1   EEAT

[†]**Warning**: Euclid's EAT is not the same as the Exterior Angle Theorem usually studied
in high school.

Look at EEAT on the plane, on a sphere, and on a hyperbolic plane.

SUGGESTIONS

You may find the following hint (which is found in Euclid's writings) useful: Draw a line from the vertex of α to the midpoint, M, of the opposite side, BC. Extend that line beyond M to a point A' in such a way that $AM \cong MA'$. Join A' to C. This hint will be referred to as *Euclid's hint*, and is pictured in Figure 8.2.

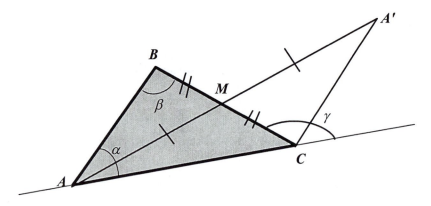

Figure 8.2   Euclid's Hint

Be cautious transferring this hint to a sphere. It will probably help to draw Euclid's hint directly on a physical sphere.

It is not necessary to use Euclid's hint to prove EEAT, and in fact many people don't "see" the hint. Another perfectly good way to prove EEAT is to use Problem **8.2**. Problems **8.1** and **8.2** are very closely related, and they can be done in either order. It is also fine to use **8.1** to prove **8.2** or use **8.2** to prove **8.1**, but of course don't do both. As a final note, remember you do not have to look at figures using only one orientation — rotations and reflections of a figure do not change its properties, so if you have trouble "seeing" something, check to see if it's something you're familiar with by orienting it differently on the page.

EEAT is not always true on a sphere, even for small triangles. Look at a counterexample as depicted in Figure 8.3. Then look at your proof of EEAT on the plane. It is very likely that your proof uses properties of angles and triangles that are true for small triangles on the sphere. Thus it may appear to you that your planar proof is also a valid proof of EEAT for small triangles on the sphere. But, there is a counterexample.

This could be, potentially, a very creative situation for you — **whenever you have a proof and counterexample of the same result, you have an opportunity to learn something deep and meaningful**. So, try out your planar proof of EEAT on the counterexample in Figure 8.3 and see what happens. Then try it on both large and small spherical triangles. If you can determine exactly which triangles satisfy EEAT and which triangles don't satisfy EEAT, then this information will be useful (but not crucial) to you in later problems.

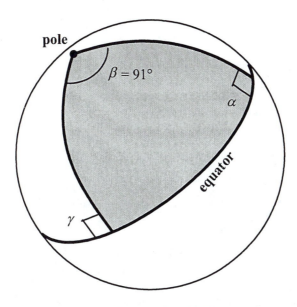

Figure 8.3   Counterexample to EEAT on a sphere

PROBLEM **8.2** SYMMETRIES OF PARALLEL TRANSPORTED LINES

*Consider two lines, **r** and **r'**, that are parallel transports of each other along a third line, **l**. Consider now the geometric figure that is formed by the three lines, one of them being a transversal to the other two, and look for the symmetries of that geometric figure.*

What can you say about the lines **r** *and* **r'***? Do they intersect? If so, where? Look at the plane, spheres, and hyperbolic planes.*

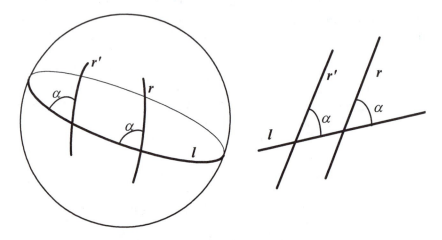

Figure 8.4a   What can you say about **r** and **r'**?

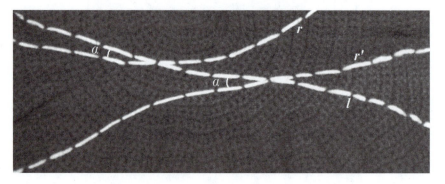

Figure 8.4b   What can you say about **r** and **r'**?

SUGGESTIONS

Parallel transport was already *informally* introduced in Chapter 7. In Problem **8.2** you have an opportunity to explore the concept further and prove its implications on the plane and a sphere. You will study the relationship between parallel transport and parallelism, as well.

A common high school definition of parallel lines is something like "two lines in a plane that never meet." But this is an inhuman definition — there is no way to check all points on both lines to see if they ever meet. This definition is also irrelevant on a sphere because we know that all geodesics on a sphere *will* cross each other. But we can measure the angles of a transversal. This is why it is more useful to talk about lines as parallel transports of one another rather than as parallel. So the question becomes:

> *If a transversal cuts two lines at congruent angles, are the lines in fact parallel in the sense of not intersecting?*

There are many ways to approach this problem. First, be sure to look at the symmetries of the local portion of the figure formed by the three lines. See what you can say about global symmetries from what you find locally. For the question of parallelism, you can use EEAT, but not if you used this problem to prove EEAT previously. Also, don't underestimate the power of symmetry when considering this problem. Many ideas that work on the plane will also be useful on a sphere and a hyperbolic plane, so try your planar proof on a sphere and a hyperbolic plane before attempting something completely different.

What is meant by *symmetry* with regard to geometric figures? A transformation is a *symmetry* of a geometric figure if it transforms that figure into itself. That is, the figure looks the same before and after the transformation. Here, we are looking for the symmetries of the plane and sphere and hyperbolic plane.

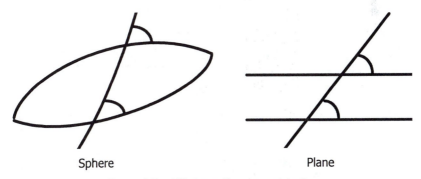

Sphere   Plane

Figure 8.5   What are the symmetries?

From Figure 8.5, we can see that on a sphere we are looking for the symmetries of a *lune* cut at congruent angles by a geodesic. A *lune* is a spherical region bounded by two half great circles.

You may be inclined to use one or both of the following results: *Any transversal of a pair of parallel lines cuts these lines at congruent angles* (Problem **10.1**). And, *the angles of any triangle add up to a straight angle* (Problem **10.4**). The use of these results should be avoided for now, as they are both false on both a sphere and a hyperbolic plane. We have been investigating what is common between the plane, spheres, hyperbolic planes — trying to use common proofs whenever possible. In addition, we will find that it is necessary to make further assumptions about the plane before we can prove these results, and no additional assumptions are needed for Problem **8.2**. You may be tempted to use other properties of parallel lines that seem familiar to you, but in each case ask yourself whether or not the property is true on a sphere and on a hyperbolic plane. If it is not true on these surfaces, then don't use it here because it is not needed.

PROBLEM 8.3   TRANSVERSALS THROUGH A MIDPOINT

a. Prove: *If two geodesics r and r' are parallel transports along another geodesic l, then they are also parallel transports along any geodesic passing through the midpoint of the segment of l between r and r'. Does this hold for the plane, spheres, and hyperbolic planes?*

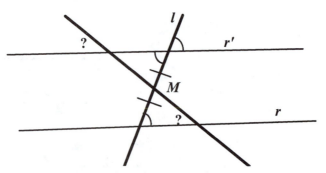

Figure 8.6 Transversals through a midpoint

b. *On a sphere or hyperbolic plane, are these the only lines that will cut r and r' at congruent angles? Why?*

c. Prove: *Two geodesics (on the plane, spheres, or hyperbolic planes) are parallel transports of each other if and only if they have a common perpendicular. On spheres and hyperbolic planes the common perpendicular is unique.*

All parts of this problem continue the ideas presented in Problem **8.2**. In fact, you may have proven this problem while working on **8.2** without even knowing it. There are many ways to approach this problem. Using symmetry is always a good way to start. You can also use some of the triangle congruence theorems that you have been working with. Look at the things you have discovered about transversals from Problems **8.1** and **8.2**; they are very applicable here. For the hyperbolic plane you may want to use results from Chapters 5 and/or 7.

PROBLEM **8.4**   WHAT IS "PARALLEL"?

Since Chapter 7, you have been dealing with issues of parallelism. Parallel transport gives you a way to check parallelism. Even though parallel transported lines intersect on the sphere, there is a *feeling of local parallelness* about them. In most applications of parallel lines the issue is not whether the lines ever intersect, but whether a transversal intersects them at congruent angles at certain points; that is, whether the lines are *parallel transports of each other along the transversal.* You may choose to avoid definitions of parallel that do not give you a direct method of verification, such as these common definitions for parallel lines in the plane:

1. *Parallel lines are lines that never intersect;*

2. *Parallel lines are lines such that any transversal cuts them at congruent angles*; or,

3. *Parallel lines are lines that are everywhere equidistant.*

a. *Check out for each of these three definitions whether they apply to parallel transported lines on a sphere or on a hyperbolic plane.*

This is closely related to Problems **8.2** and **8.3**.

b. *Show that there are pairs of geodesics on a hyperbolic plane that do not intersect and yet there are **no** transversals that cut at congruent angles. That is, the geodesics are parallel (in the sense of not intersecting) but not parallel transports of each other.*

Use the results of **8.2** and **8.3** and look on your hyperbolic plane for the boundary between geodesics that intersect and geodesics that are parallel transports.

We call a pair of geodesics that satisfy Part **b** by the name ***asymptotic geodesics***. Be warned that in many texts these geodesics are called simply *parallel*.

*c. *Show that there are pairs of geodesics on any cone with cone angle greater than 360° that do not intersect and yet there are **no** transversals that cut at congruent angles. That is, the geodesics are parallel (in the sense of not intersecting) but not parallel transports of each other along any straight (in the sense of symmetry) line.*

Experiment with a paper cone with cone angle greater than 360°.

These examples should help you realize that parallelism is not just about non-intersecting lines and that the meaning of parallel is different on different surfaces. You will explore and discuss these various notions of parallelism and parallel transport further in Chapters 9–13. Because we have so many (often unconscious) connotations and assumptions attached to the word "parallel," we find it best to avoid using the term "parallel" as much as possible in this discussion. Instead we will use terms such as "parallel transport," "non-intersecting," and "equidistant," which make explicit the meaning that is intended.

In Chapter 9, we will continue our explorations of triangle congruence theorems, some of which involve parallel transport.

In Chapter 10, we will consider various parallel postulates and explore how they apply on the plane, spheres, and hyperbolic planes. We will assume the parallel postulates on the plane and use them to prove the properties of non-intersecting and parallel transported lines on the plane. In the process, you may learn something about the history and philosophy of parallel lines and the postulates that have been used in attempts to understand parallelism.

In Chapter 11, our understanding of different notions of parallel will help us to explore isometries and patterns.

In Chapter 12, we will study parallelograms and rectangles (and their analogues on spheres and hyperbolic planes) and in the process show that, on the plane, non-intersecting lines are equidistant.

In Chapter 13, we will use the results from Chapter 11 to explore results that are only true on the plane, such as the Pythagorean Theorem and results about similar triangles.

Chapter 9

SSS, ASS, SAA, AND AAA

Things which coincide with one another are equal to one another.
> — Euclid, *Elements*, Common Notion 4 [Appendix A]

This chapter is a continuation of the triangle congruence properties studied in Chapter 6.

PROBLEM 9.1   SIDE-SIDE-SIDE (SSS)

Are two triangles congruent if the two triangles have congruent corresponding sides? Look at plane, spheres, and hyperbolic planes.

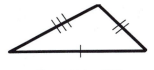

Figure 9.1   SSS

SUGGESTIONS

Start investigating SSS by making two triangles coincide as much as possible, and see what happens. For example, in Figure 9.2, if we line up one pair of corresponding sides of the triangles, we have two different orientations for the other pairs of sides as depicted in Figure 9.2. Of course, it is up to you to determine if each of these orientations is actually possible, and to prove or disprove SSS. Again, symmetry can be very useful here.

Figure 9.2   Are these possible?

On a sphere, SSS doesn't work for all triangles. The counterexample in Figure 9.3 shows that no matter how small the sides of the triangle are, SSS does not hold because the three sides always determine two different triangles on a sphere. Thus, it is necessary to restrict the size of more than just the sides in order for SSS to hold on a sphere. Whatever argument you used for the plane should work for *suitably defined* small triangles on the sphere and all triangles on a hyperbolic plane. Make sure you see what it is in your argument that doesn't work for large triangles on a sphere.

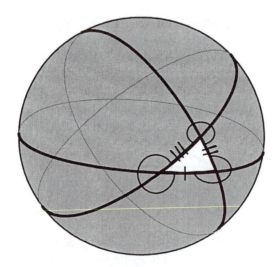

Figure 9.3   A large triangle with small sides

There are also other types of counterexamples to SSS on a sphere. Can you find them?

PROBLEM **9.2** ANGLE-SIDE-SIDE (ASS)

a. *Are two triangles congruent if an angle, an adjacent side, and the opposite side of one triangle are congruent to an angle, an adjacent side, and the opposite side of the other? Look at plane, spheres, and hyperbolic planes.*

Figure 9.4   ASS

SUGGESTIONS

Suppose you have two triangles with the above congruencies. We will call them ASS triangles. We would like to see if, in fact, the triangles are congruent. We can line up the angle and the first side, and we know the length of the second side (*BC* or *B'C'*), but we don't know where the second and third sides will meet. See Figure 9.5.

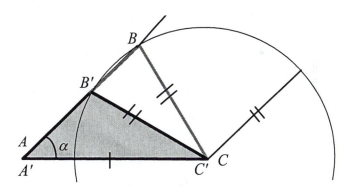

Figure 9.5   ASS is not true, in general

Here, the circle that has as its radius the second side of the triangle intersects the ray that goes from *A* along the angle *α* to *B* twice. So ASS doesn't work for all triangles on the plane or spheres or hyperbolic planes. Try this for yourself on these surfaces to see what happens. Can you make ASS work for an appropriately restricted class of triangles? On a sphere, also look at triangles with multiple right angles, and, again, define "small" triangles as necessary. Your definition of small triangle here may be very different from your definitions in Problems **6.4** and **6.5**.

There are numerous collections of triangles for which ASS is true. Explore. See what you find on all three surfaces.

> **b.** *Show that ASS holds for right triangles on the plane (where the Angle in Angle-Side-Side is right).*

This result is often called the *Right-Leg-Hypotenuse Theorem* (RLH), which can be expressed in the following way:

> **RLH:** *On the plane, if the leg and hypotenuse of one right triangle are congruent to the leg and hypotenuse of another right triangle, then the triangles are congruent.*
>
> *What happens on a sphere and a hyperbolic plane?*

At this point, you might conclude that RLH is true for small triangles on a sphere. But there *are* small triangle counterexamples to RLH on spheres! The counterexample in Figure 9.6 will help you to see some ways in which spheres are intrinsically very different from the plane. We can see that the second leg of the triangle intersects the geodesic that contains the third side an infinite number of times. So on a sphere there are small triangles that satisfy the conditions of RLH although they are non-congruent. What about RLH on a hyperbolic plane?

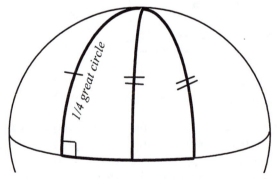

Figure 9.6   Counterexample to RLH on a sphere

However, if you look at your argument for RLH on the plane, you should be able to show:

On a sphere, RLH is valid for a triangle with all sides less than 1/4 of a great circle.

RLH is also true for a much larger collection of triangles on a sphere. Can you find such a collection? What about on a hyperbolic plane?

PROBLEM **9.3** SIDE-ANGLE-ANGLE (SAA)

Are two triangles congruent if one side, an adjacent angle, and the opposite angle of one triangle are congruent, respectively, to one side, an adjacent angle, and the opposite angle of the other triangle? Look at plane, spheres, and hyperbolic planes.

SUGGESTIONS

As a general strategy when investigating these problems, start by making the two triangles coincide as much as possible. You did this when investigating SSS and ASS. Let us try it as an initial step in our proof of SAA. Line up the first sides and the first angles. Because we don't know the length of the second side, we might end up with a picture similar to Figure 9.7.

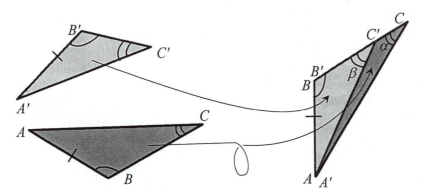

Figure 9.7   Starting SAA

The situation shown in Figure 9.7 may seem to you to be impossible. You may be asking yourself, "Can this happen?" If your temptation is to argue that α and β cannot be congruent angles and that it is not possible to construct such a figure, behold Figure 9.8.

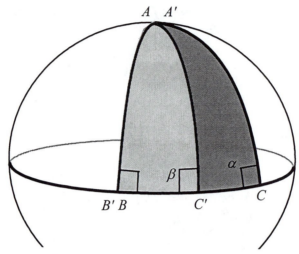

Figure 9.8   A counterexample to SAA

You may be suspicious of this example because it is not a counter-example on the plane. You may feel certain that it is the only counterexample to SAA on a sphere. In fact, we can find other counterexamples for SAA on a sphere.

With the help of parallel transport, you can construct many counterexamples for SAA on a sphere. If you look back to the first counterexample given for SAA, you can see how this problem involves parallel transport, or similarly how it involves Euclid's Exterior Angle Theorem, which we looked at in Problem **8.1**.

Can we make restrictions such that SAA *is* true on a sphere? You should be able to answer this question by using the fuller understanding of parallel transport you gained in Problems **8.1** and **8.2**. You may be tempted to use the result, *the sum of the interior angles of a triangle is 180°*, in order to prove SAA on the plane. This result will be proven later (Problem **10.1b**) for the plane, but we saw in Problem **7.3** that it does not hold on spheres and hyperbolic planes. Thus, we encourage you to avoid using it and to use the concept of parallel transport instead. This suggestion stems from our desire to see what is common between the plane and the other two surfaces, as much as possible. In addition, before we can prove that the sum of the angles of a triangle is 180°, we will have to make some additional assumptions on the plane that are not needed for SAA.

PROBLEM **9.4**   ANGLE-ANGLE-ANGLE (AAA)

Are two triangles congruent if their corresponding angles are
congruent? Look at plane, spheres, hyperbolic planes.

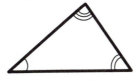

Figure 9.9   AAA

Two triangles that have corresponding angles congruent are called
similar triangles.

As with the three previous problems, make the two AAA triangles
coincide as much as possible. We know that we can line up one of the
angles, but we don't know the lengths of either of the sides coming from
this angle. So there are two possibilities: (1) Both sides of one are longer
than both sides of the other, as the example in Figure 9.10 shows on the
plane or (2) One side of the first triangle is longer than the corresponding side of the second triangle and vice versa, as the example in Figure
9.11 shows on a sphere.

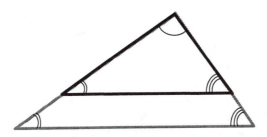

Figure 9.10   Is this possible?

As with Problem **9.3**, you may think that the example in Figure 9.11
cannot happen on a plane, a sphere, or a hyperbolic plane. The possible
existence of a counterexample relies heavily on parallel transport — you
can identify the parallel transports in each of the examples given. Try
each counterexample on the plane, on a sphere, and on a hyperbolic
plane and see what happens. If these examples are not possible, explain
why, and if they are possible, see if you can restrict the triangles sufficiently so that AAA does hold.

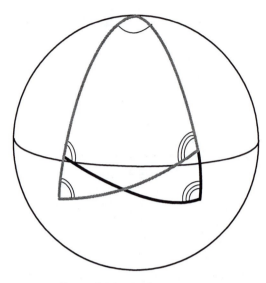

Figure 9.11   Is this possible?

Parallel transport shows up in AAA, similar to how it did in SAA, but here it happens simultaneously in two places. In this case, you will recognize that parallel transport produces similar triangles that are not necessarily congruent. However, there are no similar triangles on either a sphere or a hyperbolic plane (as you will see after you finish AAA), so you certainly need a proof that shows why such a construction is possible and why the triangles are not congruent. The construction may seem intuitively possible to you, but you should justify why it is a counterexample. Again you may need properties of parallel lines from Problem **7.3** and Problems **8.1** through **8.4**. You may also use the property of parallel transport on the plane stated in Problem **10.1** — you can assume this property now as long as you are sure not to use AAA when proving it later.

On a sphere or on a hyperbolic plane, is it possible to make the two parallel transport constructions shown in Figure 9.5 and thus get two non-congruent triangles? Try it and see. It is important that you make such constructions and that you study them on a model of a sphere and on a model of a hyperbolic plane.

Chapter 10

PARALLEL POSTULATES

[Euclid's Fifth Postulate] ought to be struck from the postulates altogether. For it is a theorem — one that invites many questions... — and requires for its demonstration a number of definitions as well as theorems. ... it lacks the special character of a postulate.

— Proclus (Greek, 410–485) [**AT**: Proclus], p. 151

PARALLEL LINES ON THE PLANE ARE SPECIAL

Up to this point we have not had to assume anything about parallel (non-intersecting) lines. No version of a parallel postulate has been necessary, on the plane, on a sphere, or on a hyperbolic plane. We defined the concrete notion of parallel transport and proved in Problem **8.2** that, on the plane (and hyperbolic planes), parallel transported lines do not intersect. Now in this chapter we will look at three important properties on the plane that require further assumptions and that will be needed in later chapters. If you are willing to assume these statements you may skip this chapter but we urge you to at least finish reading this first section.

If two lines on the plane are parallel transports of each other along some transversal, then they are parallel transports along any transversal. (Problem **10.1**)

On the plane, the sum of the interior angles of a triangle is always 180°. (Problem **7.3b** or Problem **10.2b**)

On the plane, non-intersecting lines are parallel transports along all transversals. (Problem **10.3d**)

We have already seen that none of these properties are true on a sphere or a hyperbolic plane. Thus all three need for their proofs some property of the plane that does not hold on spheres and hyperbolic planes. The various properties that permit proofs of these three statements are collectively termed the ***Parallel Postulates***.

Only the three statements above are needed from this chapter for the rest of the book. Therefore, it is possible to omit this chapter and assume one of the above three statements and then prove the others. However, parallel postulates have a historical importance and a central position in many geometry textbooks and in many expositions about non-Euclidean geometries. The problems in this chapter are an attempt to help people unravel and enhance their understanding of parallel postulates. Comparing situations on the plane with situations on a sphere and on a hyperbolic plane is a powerful tool for unearthing our hidden assumptions and misconceptions about the notion of "parallel" on the plane.

Because we have so many (often unconscious) connotations and assumptions attached to the word "parallel," we find it best to avoid using the term "parallel" as much as possible in this discussion. Instead we will use terms such as "parallel transport," "non intersecting," and "equidistant," which make explicit the meaning that is intended.

PROBLEM 10.1   PARALLEL TRANSPORT ON THE PLANE

*Show that if l_1 and l_2 are lines on the plane such that they are parallel transports along a transversal l, then they are parallel transports along any transversal. Prove this using any assumptions you find necessary. Make as few assumptions as you can, and make them as simple as possible. **Be sure to state your assumptions clearly.***

What part of your proof does not work on a sphere or on a hyperbolic plane?

SUGGESTIONS

This problem is by no means as trivial as it at first may appear. In order to prove this theorem, you will have to assume something — there are many possible assumptions, so use your imagination. But at the same time, try not to assume any more than is necessary. If you are having trouble deciding what to assume, try to solve the problem in a way that seems natural to you and then see what develops, while making explicit any assumptions you are using.

On spheres and hyperbolic planes, try the same construction and proof you used for the plane. What happens? You should find that your proof does not work on these surfaces. So, what is it about your proof (on a sphere and hyperbolic plane) that creates difficulties?

Problem **10.1** emphasizes the differences between parallelism on the plane and parallelism on spheres and hyperbolic planes. On the plane, non-intersecting lines exist, and one can "parallel transport" everywhere. Yet, as was seen in Problems **8.2** and **8.3**, on spheres and hyperbolic planes two lines are cut at congruent angles if and only if the transversal line goes through the center of symmetry formed by the two lines. That is, on spheres and hyperbolic planes two lines are parallel transports only when they can be parallel transported through the center of symmetry formed by them. Be sure to draw a picture locating the center of symmetry and the transversal. On spheres and hyperbolic planes it is impossible to slide the transversal along two parallel transported lines while keeping both angles constant (something you *can* do on the plane). In Figures 10.1a and 10.1b, the line r' is a parallel transport of line r along line l, but it is not a parallel transport of r along l'.

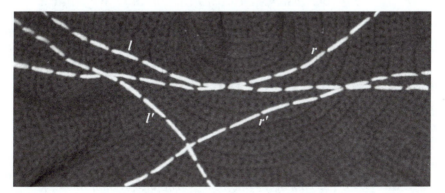

Figure 10.1a   Parallel transport on a hyperbolic plane along *l* but not along *l'*

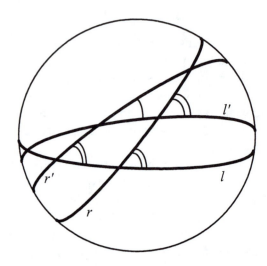

Figure 10.1b   Parallel transport on a sphere along *l*, but not along *l'*

Pause, explore, and write out your ideas before reading further.

We will now divide parallel postulates into three groups: those involving mostly parallel transport, those involving mostly equidistance, and those involving mostly intersecting or non-intersecting lines. This division is useful even though it is rough and unlikely to fit every conceivable parallel postulate. *Which of these three groups is the most appropriate for your assumption from Problem* **10.1**? We will call the assumption you made in Problem **10.1** "*your parallel postulate*".

PROBLEM 10.2 PARALLEL POSTULATES NOT INVOLVING (NON-) INTERSECTING LINES

Commonly used postulates of this sort are

H = 0:   *The holonomy of triangles is zero.*

A = 180:   *The sum of the angles of a triangle is equal to 180°.*

> **PT!**:   *If two lines are parallel transports (PT) along one line*
> *then they are PT along ALL transversals.*

Note that these are false on spheres and on hyperbolic planes. The last two properties (**A = 180** and **PT!**) will be needed crucially in almost all the remaining chapters (and you may have already used one of them in Problem **9.4**). These properties are needed to study (on the plane) parallelograms, rectangles, and similar triangles (Chapters 12–13), circles (Chapter 14), projections of spheres and hyperbolic planes onto the Euclidean plane (Chapters 15–16), Euclidean manifolds (Chapters 17 and 22), solutions to quadratic and cubic equations (Chapter 18), trigonometry (Chapter 19), and polyhedra in 3-space (Chapter 21). It is these properties and their uses that we most often associate with consequences of the parallel postulates.

In Problem **7.3b** we proved **A = 180** on the plane, so we can

a.   *Ask what assumption about the plane was used in proving **7.3b**?*

b.   *Use **A = 180** to prove **PT!**.*

Look first at transversals that intersect the line along which the two lines are parallel transports.

c.   *Using **PT!** (Problem **10.1**) prove **A = 180** without using results from Chapter 7.*

Start by parallel transporting one side of the triangle.

d.   *Show that, on the plane, **H = 0** ⇔ **A = 180** ⇔ **PT!**. If your postulate from **10.1** is in this group then show that it is also equivalent to the others.*

This is usually accomplished most efficiently by proving **H = 0** ⇒ **A = 180** ⇒ **PT!** ⇒ Your Postulate ⇒ **H = 0**, or in any other order.

e.   *Prove that, on the plane, two parallel transported lines are equidistant.*

Look for rectangles or parallelograms.

EQUIDISTANT CURVES ON SPHERES
AND HYPERBOLIC PLANES

The latitude circles on the earth are sometimes called "Parallels of Latitude." They are parallel in the sense that they are everywhere equidistant as are concentric circles on the plane. In general, transversals do not cut equidistant circles at congruent angles. However, there is one important case where transversals do cut the circles at congruent angles. Let l and l' be latitude circles the same distance from the equator on opposite sides of it. See Figure 10.2. Then, every point on the equator is a center of half-turn symmetry for these pair of latitudes. Thus, as in Problems **8.3** and **10.1**, every transversal cuts these latitude circles in congruent angles.

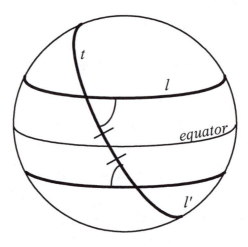

Figure 10.2   Special equidistant circles

The same ideas work on a hyperbolic plane: If g is a geodesic and l and l' are the (two) curves that are a distance d from g, then l and l' are equidistant from each other and every transversal cuts them at congruent angles. This follows from the fact that g has half-turn symmetry at every point.

PROBLEM **10.3** PARALLEL POSTULATES INVOLVING (NON-) INTERSECTING LINES

One of Euclid's assumptions constitutes *Euclid's Fifth* (*or Parallel*) *Postulate* (*EFP*), which says

EFP: *If a straight line intersecting two straight lines makes the interior angles on the same side less than two right angles, then the two lines (if extended indefinitely) will meet on that side on which are the angles less than two right angles.*

For a picture of EFP, see Figure 10.3.

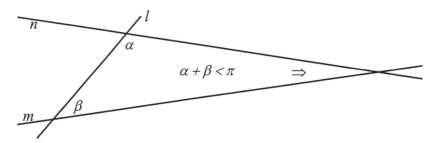

Figure 10.3   Euclid's Parallel Postulate

You probably did not assume EFP in your proof of Problem **10.1**. You are in good company — many mathematicians, including Euclid, have tried to avoid using it as much as possible. However, we will explore EFP because, historically, it is important, and because it has some very interesting properties as you will see in Problem **10.3**. On a sphere, all straight lines intersect twice, which means that EFP is trivially true on a sphere. But in Problem **10.3**, you will show that EFP is also true in a stronger sense on spheres. You will also be able to prove that EFP is false on a hyperbolic plane.

Thus, EFP does not have to be assumed on a sphere — it can be proved! However, in most geometry textbooks, EFP is replaced by another postulate, claimed to be equivalent to EFP. This postulate is

Playfair's Parallel Postulate (*PPP*), and it can be expressed in the following way:

PPP: *For every line l and every point P not on l, there is a unique line l' which passes through P and does not intersect (is parallel to) l.*

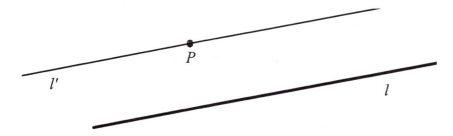

Figure 10.4   Playfair's Parallel Postulate

Note that, because any two great circles on a sphere intersect, there are *no* lines *l'* that are parallel to *l* in the "not intersecting" sense. Therefore, Playfair's Postulate is not true on spheres. On the other hand, if we change "parallel" to "parallel transport" then *every* great circle through *P* is a parallel transport of *l* along *some* transversal. What happens on a hyperbolic plane? In Problem **10.3**, you will explore the relationships among EFP, Playfair's Postulate, and your postulate from Problem **10.1**. In Problem **10.4** we will explore these postulates on spheres and hyperbolic planes.

   a. *Show that, on the plane, **EFP** and **PPP** are equivalent. If your postulate from **10.1** is in this group, is it equivalent to the others? Why or why not?*

To show that **EFP** and **PPP** are equivalent on the plane, you need to show that you can prove **EFP** if you assume **PPP** and vice versa. If the three postulates are equivalent, then you can prove the equivalence by showing that

$$\textbf{EFP} \Rightarrow \textbf{PPP} \Rightarrow \text{Your Postulate} \Rightarrow \textbf{EFP}$$

or in any other order. It will probably help you to draw lots of pictures of what is going on. Also, remember that we proved in Problem **8.2** (without using any parallel postulate) that parallel (non-intersecting) lines exist. Note that **PPP** is not true on a sphere but **EFP** is true, so your proof that **EFP** implies **PPP** on the plane must use some property of the plane that does not hold on a sphere. Look for it.

b. *Prove that either* **EFP** *or* **PPP** *can be used to prove* (*without using* **10.1** *or* **7.3b**) *one of* **H** = **0**, **A** = **180**, *or* **PT!**. (*It does not matter which.* *Why?*)

So, are all these postulates equivalent to each other on the plane? The answer is almost, but not quite! In order for **A** = **180** (or **H** = **0** or **PT!**) to imply **EFT** (or **PPP**), we have to make an additional assumption, the ***Archimedean Postulate*** (in some books this is called the *Axiom of Continuity*), named after the Greek mathematician, Archimedes (who lived in Sicily, 287?–212 B.C.):

*AP: *On a line, if the segment AB is less than* (contained in) *the segment AC, then there is a finite* (positive) *integer, n, such that, if we put n copies of AB end-to-end* (see Figure 10.5), *then the n-th copy will contain the point C.*

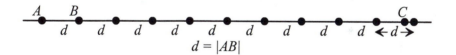

$$d = |AB|$$

Figure 10.5   The Archimedean Postulate

The Archimedean Postulate can also be interpreted to rule out the existence of infinitesimal lengths. The reason these postulates are "almost but not quite" equivalent to each other on the plane is that though **AP** is needed it is assumed by most people to be true on the plane, spheres, and hyperbolic planes. But this is the first time we have needed this assumption in this book.

*c. *Show that, on the plane,* **AP** *and* **A** = **180** *imply* **EFP**.

Look at the situation of **EFP**, which we redraw in Figure 10.6. Pick a sequence of equally spaced points *A*, *A'*, *A''*,... on *n* (on the side of the angle *α*). Next parallel transport *l* along *m* to lines *l'*, *l''*,... that intersect *n* at the points *A'*, *A''*,... . And parallel transport *m* along *l* to *m'*, *m''*,... which also intersect *n* at the points *A'*, *A''*,... . Look for congruent angles and congruent triangles. Use **AP** to argue that *n* and *l* will eventually intersect.

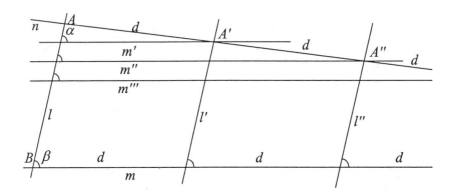

Figure 10.6   **AP** and **A = 180** imply **EFP**

Clearly, this proof feels different from the others proofs in this chapter. And, as already pointed out, most of the applications in this book of parallel postulates only needs **A = 180** and **PT!**. So is it possible for us not to bother with **EFP** and **PPP**? Or are they needed? Yes, we need them, but only in a few places such as in Problem **11.4** where what we need to know is

d. Prove using **EFP**: *On the plane, non-intersecting lines are parallel transports along every transversal.*

Compare with **EFP**.

PROBLEM 10.4 EFP AND PPP ON SPHERE AND HYPERBOLIC PLANE

a. *Show that EFP is true on a sphere in a strong sense; that is, if lines l and l' are cut by a transversal t such that the sum of the interior angles α + β on one side is less than two right angles, then not only do l and l' intersect, but they also intersect "closest" to t on the side of α and β. You will have to determine an appropriate meaning for "closest."*

To help visualize the postulates, draw these "parallels" on an actual sphere. There are really two parts in this proof — first, you must come up with a definition of "closest," then, prove that EFP is true for this definition. The two parts may come about simultaneously as you come up with a proof. This problem is closely related to Euclid's Exterior Angle Theorem, but can also be proved without using EEAT. One case that you should look at specifically is pictured in Figure 10.7. It is not necessarily obvious how to define the "closest" intersection.

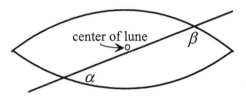

Figure 10.7   Is EFP true on a sphere?

b. *On a hyperbolic plane let **l** be a geodesic and let **P** be a point not on **l**, then show that there is an angle θ with the property that any line **l′** passing through **P** is parallel to* (not intersecting) *l if the line **l′** does not form an angle less than θ with the line from **P** which is perpendicular to l* (Figure 10.8).

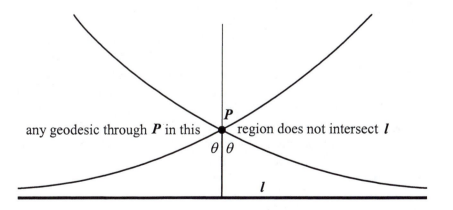

Figure 10.8   Multiple parallels on a hyperbolic plane

Look at a variable line through P that does intersect l and then look at what happens as you move the intersection point out to infinity.

> **c.** *Using the notion of parallel transport, change Playfair's Postulate so that the changed postulate is true on both spheres and hyperbolic planes. Make as few alterations as possible and keep some form of uniqueness.*

Try to limit your alterations so that the new postulate preserves the spirit of the old one. You can draw ideas from any of the previous problems to obtain suitable modifications. Prove that your modified versions of the postulate are true on spheres and hyperbolic planes.

> **d.** *Either prove your postulate from Problem **10.1** on a sphere and on a hyperbolic plane or change it, with as few alterations as possible, so that it is true on these surfaces. You may need to make different changes for the two surfaces.*

In Problem **10.1** you should have decided whether or not your postulate is true on spheres or on hyperbolic spaces.

COMPARISONS OF PLANE, SPHERES, AND HYPERBOLIC PLANES

Figure 10.9 is an attempt to represent the relationships among parallel transport, non-intersecting lines, EFP, and Playfair's Postulate. Can you fit your postulate into the diagram?

Playfair's Postulate assumes both the existence and uniqueness of parallel lines. In Problem **8.2**, you proved that if one line is a parallel transport of another, then the lines do not intersect on the plane or on a hyperbolic plane. Thus, it is not necessary to assume the existence of parallel lines (non-intersecting lines) on the plane or hyperbolic plane.

On a sphere any two lines intersect. However, in Problem **8.4** we saw that there *are* non-intersecting lines that are not parallel transports of each other on a hyperbolic plane and on any cone with cone angle larger than 360°.

	Euclidean	Hyperbolic	Spherical
parallel transported lines	do not intersect and are equidistant **8.2 & 10.2**	diverge in both directions **8.2 & 8.3**	always intersect **2.1**
parallel transported lines	are parallel transports along all transversals **10.1**	are parallel transports along any transversal that passes through the center of symmetry of the lines **8.3**	
non-intersecting lines	are parallel transports along all transversals **10.3d**	sometimes are not parallel transports **8.4b**	do not exist **2.1**
Euclid's 5th Postulate	must be assumed Chapter 10	is false **8.4b**	is provable in a strong sense **10.4a**
Playfair's Parallel Postulate	a unique line through a point not intersecting a given line **10.3**	many lines through a point not intersecting a given line **10.4b**	no non-intersecting lines **2.1**
your postulate from **10.1**			
two points determine	a unique line and line segment **6.1d**		at least two line segments **6.1d**
sum of the angles of a triangle	$= 180°$ **7.3b & 10.2**	$< 180°$ **7.1**	$> 180°$ **7.2**
holonomy	$= 0°$ **10.2**	$< 0°$ **7.4**	$> 0°$ **7.4**
VAT & ITT	are always true, **3.1 & 6.2**		
SAS, ASA, and SSS	hold for all triangles **6.4, 6.5, & 9.1**		hold for small triangles **6.4, 6.5, & 9.1**
AAA	is false, **9.4**, similar triangles	is true, **9.4**, no similar triangles	

Figure 10.9   Comparisons of the three geometries

SOME HISTORICAL NOTES ON THE PARALLEL POSTULATES

The problem of parallels puzzled Greek geometers (see the quote at the beginning of this chapter). The prevailing belief was that it surely follows from the straightness of lines that lines like n and m in Figure 10.10 *had* to intersect, and so Euclid's Fifth Postulate would turn out to be unnecessary.

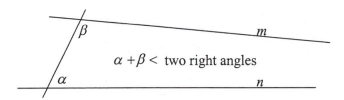

$$\alpha + \beta < \text{ two right angles}$$

Figure 10.10   Surely these intersect!

During and since the Greek era, there have been attempts to derive the postulate from the rest of elementary geometry; attempts to reformulate the postulate or the definition of parallels into something less objectionable; and descriptions of what geometry would be like if the postulate was in some way denied. Part of these attempts was to study what geometric properties could be proved without using a parallel postulate, but with the other properties of the plane. The results of these studies became known as ***absolute geometry***. Both the plane and a hyperbolic plane are examples of absolute geometry. Included in absolute geometry are all the results about the plane and hyperbolic planes that we have discussed in Chapters 1, 3, 5, 6, 8, and 9 (except AAA). In general, absolute geometry includes everything that is true of both the plane and hyperbolic planes.

Claudius Ptolemy (Greek, 100–178) implicitly use what we now call PPP to prove EFP. This is reported to us by Proclus (Greek, 410–485) [**AT:** Proclus] who explicitly states PPP. Playfair's Parallel Postulate got its current name from the Scottish mathematician, John Playfair (1748–1819), who brought out successful editions of Euclid's *Elements* in the years following 1795. After 1800 many commentators referred to Playfair's postulate (PPP) as the best statement of Euclid's postulate, so it became a tradition in many geometry books to use PPP instead of EFP. In absolute geometry PPP is equivalent to EFP (as you show in **10.3a**), and because absolute geometry was the focus of many investigations the statement "PPP is equivalent to EFP" was made and is still being repeated in many textbooks and expository writings about geometry even when the context is not absolute geometry. As you showed in **10.4a**, EFP is true on spheres and so cannot be equivalent to PPP, which is clearly false.

During the 9th through 12th centuries, parallel postulates were explored by mathematicians in the Islamic world. al-Hasan ibn al-Haytham (Persian, 965–1040) proved EFP by assuming: *A quadrilateral with three right angles must have all right angles.* Quadrilaterals with three right angles are known later in Western literature as Lambert Quadrilaterals after Johann Lambert (German–Swiss, 1728–1777) who studied them and also was the first to extensively investigate hyperbolic trigonometric functions. The Persian poet and geometer Omar Khayyam (1048–1131) wrote a book that in translation is entitled *Discussion of Difficulties in Euclid* [**AT:** Khayyam (1961)]. Khayyam introduces a new postulate (which he attributes to Aristotle, though it has not been found among the surviving works of Aristotle) which says: *Two straight lines which start to converge continue to converge.* In his work on parallel lines he studied the Khayyam Quadrilaterals (later in the West to be called Saccheri Quadrilaterals) that are discussed here in Chapter 12. Nassir al-Din Al-Tusi (Khorasan, 1201–1274) furthered the study of parallel postulates and is credited (though some say it may have been his son) with first proving: **A = 180** *is equivalent to* **EFP**, our **10.3bc**. Al-Tusi's works were the first Islamic mathematical works to be discovered in the Western Renaissance and were published in Rome in 1594.

The assumption *parallel lines are equidistant* (see our **10.2d**) was discussed is various forms by Aristotle (Greek, 383–322 B.C.), Posidonius (Greek, 135–51 B.C.), Proclus, ibn Sina (Uzbek, 980–1037), Omar Khayyam, and Saccheri (Italian, 1667–1733).

John Wallis (English, 1616–1703) proved that EFP was equivalent to: *To every triangle, there exists a similar triangle of arbitrary magnitude.* (See our **9.4.**). The work, already mentioned, of Wallis, Saccheri, Lambert in the 17th and 18th centuries was continued by the French school into the early 19th century by Joseph Fourier (French, 1768–1830), who concluded that geometry was a physical science and could not be established a priori, and Adrien-Marie Legendre (1752–1833), who proved that *in absolute geometry the sum of the interior angles of a triangle is always less than or equal to 180°.*

The breakthrough in the study of parallel postulates came in the 19th century when, apparently independently, C.F. Gauss (German,

1777–1855), János Bolyai (Hungarian, 1802–1860), and N.I. Lobatchevsky (Russian, 1792–1856) finally developed hyperbolic geometry as an absolute geometry that did not satisfy EFP or PPP. See Chapters 5 and 16 for more discussions of hyperbolic geometry. These discoveries of hyperbolic geometry showed that the quest for a proof of EFP from within absolute geometry is impossible.

For details of the relevant histories, see [**Hi:** Gray], [**DG:** McCleary], [**Hi:** Rosenfeld], and the Heath's editorial notes in [**AT:** Euclid, *Elements*].

Because of this long history of investigation into parallel postulates, many books misleadingly call hyperbolic geometry *the* non-Euclidean geometry. As pointed out in the introduction to Chapter 5, spherical geometry has been studied since ancient times but it did not fit into absolute geometry and thus was (and still is) left out of many discussions of non-Euclidean geometry. In the discussions that do include spherical geometry it is called by various names: Riemannian, projective, elliptic, double elliptic, and spherical. These different labels for spherical usually imply different settings and contexts as summarized in the following:

Bernhard Riemann (German, 1826–1866) pioneered the intrinsic (and analytic) view for surfaces and space; and, in particular, he introduced an intrinsic analytic view of the sphere that became know as the ***Riemann Sphere***. The Riemann Sphere is usually studied in a course on complex analysis.

Elliptic geometry usually indicates an axiomatic formalization of spherical geometry where each pair of antipodal points is identified as one point.

Double-elliptic geometry usually indicates an axiomatic formalization of spherical geometry where antipodal points are considered as two points.

Projective geometry comes from the image onto the plane of the spherical geometry (with antipodal points identified) under gnomic projection. (See Problems **15.1** and **19.6**.)

Spherical geometry usually indicates the geometry of the sphere sitting extrinsically in Euclidean 3-space.

Chapter 11

ISOMETRIES AND PATTERNS

> All shapelessness whose kind admits of pattern and
> form, as long as it remains outside of Reason and Idea, is ugly
> by that very isolation from the Divine Reason-Principle. And
> this is the Absolute Ugly: an ugly thing is something that has
> not been entirely mastered by pattern, that is by Reason, the
> Matter not yielding at all points and in all respects to
> Ideal-Form.
>
> — Plotinus, *The Enneads*, I.6.2 [**AT**: Plotinus]

In this chapter we will show for the plane, spheres, and hyperbolic
planes that every isometry is the product of reflections, and determine
the different types of isometries. This finishes the study of reflections
and rotations we started in Problem **5.3**. Then we will study patterns in
these three spaces. Along the way we will look at some group theory
through its origins, that is, geometrically.

It would be good for the reader to start by reviewing Problem **5.3**.
We will start the chapter with a further investigation of isometries and
then with a discussion of definitions and terminology. We advise you to
investigate this introductory material as concretely as possible.

PROBLEM 11.1 ISOMETRIES

Definitions:
*An **isometry** of X is a transformation that takes X onto X
and preserves all distances and angles in X.*

*A **reflection through the line (geodesic)** l is an isometry* R_l
such that it fixes only *those points that lie on l and, for each*

133

point P not on l, l is the perpendicular bisector of the geodesic segment joining P to R$_l$(P).

The existence of reflections through any geodesic is part of our notion of geodesic in the plane, spheres, and hyperbolic planes.

*A **rotation about P through the angle** θ is an isometry S$_\theta$ that leaves the point P fixed and is such that for every Q ≠ P the angle ∠QPS$_\theta$(Q) = θ.*

a. Prove: *If P is a point on the plane, sphere, or hyperbolic plane, and ∠APB is any angle at P, then there is a rotation about P through the angle θ = ∠APB.*

Refer to Problem **5.3** and Figure 5.13. Remember to show that the composition of these two reflections has the desired angle property.

But it is still not clear that anything we would want to call a translation exists on spheres or hyperbolic planes.

b. *Let m and n be two geodesics on the plane, a sphere, or a hyperbolic plane with a common perpendicular l. Look at the composition of the reflection R$_m$ through m with the reflection R$_n$ through n. Show why this composition R$_n$R$_m$ could be called a translation of the surface along l. How far are points on l moved?*

Let Q be an arbitrary point on *l* (but not on *m* or *n*). Investigate where Q is sent by R$_m$ and then by R$_n$R$_m$.

We can now make a definition that works on all surfaces:

*A **translation of distance d along the line** (**geodesic**) l is an isometry T$_d$ that takes each point on l to a point on l at the distance (along l) of d and takes each point not on l to another point on the same side of l.*

c. Prove: *If l is a geodesic on the plane, a sphere, or a hyperbolic plane, and d a distance, then there is a translation of distance d along l.*

In Figure 11.1 (which can be considered to be on either the plane, a sphere, or a hyperbolic plane), two congruent geometric figures, $\mathcal{F}$ and $\mathcal{G}$, are given, but there is not a single reflection or rotation or translation that will take one onto the other.

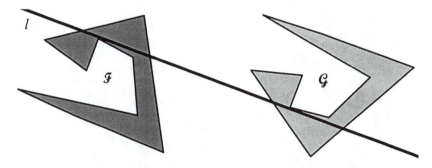

Figure 11.1   Glide reflection

However, it is clear that there is some composition of translations, rotations, and reflections that will take $\mathcal{F}$ onto $\mathcal{G}$. In fact, the composition of a reflection through the line l and a translation along l will take $\mathcal{F}$ onto $\mathcal{G}$. This composition is called *a glide reflection along l*.

> *A glide reflection* (or just plain *glide*) *of distance d along the line* (*geodesic*) *l is an isometry* G_d *that takes each point on l to a point on l at the distance* (*along l*) *of d and takes each point not on l to another point on the* other side *of l.*

   d. *If l is a geodesic on the plane, sphere, or hyperbolic plane and if d is a distance, then there is a glide of distance d along l.*

Later in this chapter we will show that these are the only isometries of the plane and spheres, and almost the only isometries on hyperbolic planes.

On the hyperbolic plane there are some pairs of geodesics, called **asymptotic geodesics**, that do not intersect and also do not have a common perpendicular (and thus are not parallel transports). See Problem **8.4b**. For example, two radial geodesics in the annular hyperbolic plane are asymptotic. The composition of reflections about two asymptotic geodesics is defined to be a **horolation**. See Figure 11.2, where

$A' = R_m(A)$, $A'' = R_n(R_m(A))$, $B'' = R_n(R_m (B))$, $C'' = R_n(R_m(C))$, and a is an annular line (not a geodesic).

Figure 11.2   Horolation

For the horolation depicted in Figure 11.2 the two reflection lines are radial geodesics. This is not really the special case it looks to be: If l and m are any two asymptotic (but not radial) geodesics then l must intersect a radial geodesic r, in fact, infinitely many radial geodesics. Reflect the whole hyperbolic plane through the bisector b of the angle between the end of l at which it is asymptotic to m and the end of r at which it is asymptotic to other radial geodesics. See Figure 11.3. The images of l and m under the reflection are now radial geodesics, $r = R_b(l)$ and $R_b(m)$.

A horolation is an isometry that is neither a rotation nor a translation, but can be thought of as a rotation about the point at infinity where the two asymptotic lines converge. In the case of radial geodesics, a horolation will take each annulus to itself (because the radial geodesics are perpendicular to the annuli. See Figure 11.2. In Chapter 16 we will also see that a horolation corresponds to a translation parallel to the x-axis.

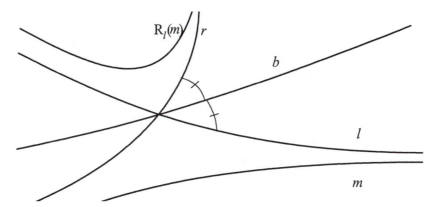

Figure 11.3   Asymptotic geodesics can be reflected to radial geodesics

SYMMETRIES AND PATTERNS

In Chapter 1 we talked about symmetries of the line. All of those symmetries can be seen as isometries of the plane except for similarity symmetry and 3-D rotation symmetry (through any angle not an integer multiple of 180°). Similarity symmetry changes lengths between points of the geometric figure and thus is not an isometry. Three-dimensional rotation symmetry is an isometry of 3-space, but it moves any plane off itself and thus cannot be an isometry of a plane (unless the angle of rotation is a multiple of 180°).

What is a symmetry of a geometric figure? A *symmetry* of a geometric figure is an isometry that takes the figure onto itself. For example, reflection through any median is a symmetry of an equilateral triangle. See Figure 11.4.

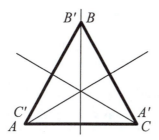

Figure 11.4   Reflection symmetries

It is easy to see that rotations through 1/3 and 2/3 of a revolution, $S_{1/3}$ and $S_{2/3}$, are also symmetries of the equilateral triangle. In addition, the identity, Id, is (trivially, by the definition) an isometry and, thus, is also a symmetry of the equilateral triangle. Therefore, the equilateral triangle has 6 symmetries: R_A, R_B, R_C, $S_{1/3}$, $S_{2/3}$, Id, where R_A, R_B, R_C, denote the reflections through the medians from A, B, C. These are the only symmetries of the equilateral triangle.

Now look at the geometric figure in Figure 11.5. It has exactly the same symmetries as an equilateral triangle. Though the two figures look very different we say they are *isomorphic patterns*.

Figure 11.5   Same symmetries as the equilateral triangle

A *pattern* is a figure *together with all its symmetries*; and we call the collection of all symmetries of a geometric figure its **symmetry group**. We want some way to denote that two different patterns have the same symmetries, as is the case with the pattern in Figure 11.5 and the equilateral triangle. We do this by saying that two patterns are *isomorphic* if they have the same symmetries. This should become clearer through more examples.

The letters S and N each have only half-turn and the identity as symmetries, thus we say they are isomorphic patterns with symmetry group $\{Id, S_{1/2}\}$. Similarly, the letters A and M each have only reflection and the identity as symmetries and thus are isomorphic patterns with symmetry group $\{Id, R\}$. We do **not** say that the letters S and A each have the same number of patterns, but we do call them isomorphic patterns because the symmetries are different symmetries. The reader should check their understanding by finding isomorphic patterns among the other letters of the alphabet.

To construct further examples, one can start with a geometric figure, often called a ***motif***, that has no symmetry (except the identity). For example, the geometric figures in Figure 11.6 are possible motifs.

Figure 11.6   Motifs

To make examples of patterns with a specific isometry we can start with any motif and then use the isometry and its inverse to make additional copies of the motif over and over again. In the process we obtain another geometric figure for which the initial isometry is a symmetry. Let us look at an example: If we start with the first motif in Figure 11.6 and the isometry is translation to the right through a distance d, then, using this isometry and its inverse (translation to the left through a distance d) and repeating them over and over, we obtain the pattern in Figure 11.7.

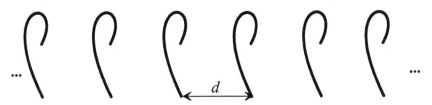

Figure 11.7   Pattern with translation symmetries

The symmetry group of the pattern in Figure 11.7 is

$$\{\text{Id (the identity)}, \text{T}_{nd} \text{ (where } n = \pm1, \pm2, \pm3,...)\}.$$

If the isometry is clockwise rotation through 1/3 of a revolution about the lower endpoint of the motif, then we obtain the pattern in Figure 11.8.

Figure 11.8   Rotation symmetry

This figure is a pattern with symmetries $\{Id, S_{1/3}, S_{2/3}\}$. Note that this pattern is non-isomorphic with the equilateral triangle pattern.

If, in the constructions depicted in Figures 11.7 and 11.8, we replace the motif with any other motif (with no non-trivial symmetries), then you will get other patterns that are isomorphic to the original ones, because the symmetries are the same.

We call the collection of all symmetries of a geometric figure its **symmetry group**. If g, h are symmetries of a figure $\mathcal{F}$, then you can easily see that

*The **composition** gh (first transform by h and then follow it by g) is also a symmetry. For example, for an equilateral triangle the composition of the reflection R_A with the reflection R_B is a rotation, $S_{2/3}$. In symbols, $R_B R_A = S_{2/3}$.*

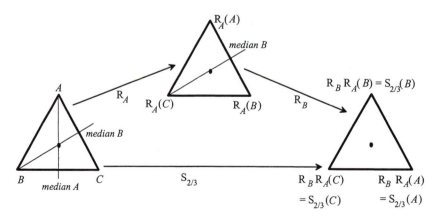

Figure 11.9   Composition of symmetries

*The **identity** transformation,* Id *(the transformation that takes every point to itself), is a symmetry.*

*For every symmetry g of a figure there is another symmetry f such that gf and fg are the identity — in this case we call f the **inverse** of g. For example, the inverse of the rotation* $S_{2/3}$ *is the rotation* $S_{1/3}$ *and vice versa. In symbols,* $S_{1/3}S_{2/3} = S_{2/3}S_{1/3} = $ Id.

Those readers who are familiar with algebraic groups will recognize the above as the axioms for a group. We will discuss further connections with group theory at the end of this chapter.[†]
A ***strip*** (or linear, or frieze) ***pattern*** is a pattern that has a translation symmetry, with all of its symmetries also symmetries of a given line. For example, the pattern in Figure 11.7 is a strip pattern as is the pattern in Figure 11.10. The strip pattern in Figure 11.10 has symmetry group: $\{$Id , R_l , T_{nd} (where $n = 0, \pm1, \pm2, \pm3, \cdots)\}$.

Figure 11.10   A strip pattern

You are now able to start to study properties about patterns and isometries.

PROBLEM 11.2 EXAMPLES OF PATTERNS

a. *Go through all the letters of the alphabet* (in normal printing by hand) *and decide which are isomorphic as patterns.*

b. *Find as many (non-isomorphic) patterns as you can which have only finitely many symmetries. List all the symmetries of each pattern you find.*

[†] If two patterns are isomorphic, then their symmetry groups are isomorphic as abstract groups. The converse is often, but not always, true. For example, the symmetry groups of the letter A and the letter S are isomorphic as abstract groups to Z_2, but they are not isomorphic as patterns because they have different symmetries.

c. *Find as many (non-isomorphic) strip patterns as you can. List all the symmetries of each strip pattern you find.*

The purpose of this problem is to get you looking at and thinking about patterns. Examples of different strip patterns and many finite patterns can be found on buildings everywhere: houses of worship, courthouses, and most older buildings. Also look for other decorations on plates or on wallpaper edging. Until we have explored isometries further in Problems **11.3** through **11.6**, you may not be able to find all non-isomorphic patterns and definitely you will not be able to prove that you have all of them.

In order to determine if your list of strip patterns and finite patterns contains all possible strip and finite patterns, we need first to explore properties of isometries. Here is one very important property of isometries of a plane: If two isometries act on three non-collinear points of the plane in the same way, then they are the same isometry. Let us see an example of how this works before you see why this property holds. In Figure 11.11, H_p represents the half-turn around p, R_m represents the reflection through line m, G_l represents a glide reflection along l, and

$$\mathcal{F}_1 = \{A, B, C\}, \quad \mathcal{F}_2 = \{A', B', C'\} \text{ and } \mathcal{F}_3 = \{A'', B'', C''\}.$$

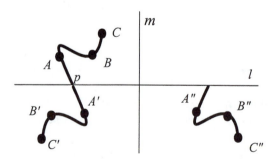

Figure 11.11   A glide equals a half-turn followed by a reflection

We can see that $H_p(\mathcal{F}_1) = \mathcal{F}_2$, $R_m(\mathcal{F}_2) = \mathcal{F}_3$, and $G_l(\mathcal{F}_1) = \mathcal{F}_3$. But then G_l and $R_m H_p$ perform the same action on the three points. If we use Problem **11.3**, we can say that $G_l = R_m H_p$; that is, $G_l(X) = R_m H_p(X)$, for all points X on the plane. You can see now the usefulness of proving Problem **11.3**.

PROBLEM **11.3**   ISOMETRY DETERMINED BY THREE POINTS

> *Prove the following: On the plane, spheres, or hyperbolic planes, if f and g are isometries and A, B, C are three non-collinear points, such that f(A) = g(A), f(B) = g(B), and f(C) = g(C), then f and g are the same isometry, that is, f(X) = g(X) for every point X.*

If you have trouble getting started with this problem, then take a specific example such as the two congruent figures, $\mathcal{F}$ and $\mathcal{G}$, at the start of this chapter and label three non-collinear points, A, B, C. Pick another point X and convince yourself as concretely as possible that any sequence of isometries that takes $\mathcal{F}$ onto $\mathcal{G}$ must always take X to the same location.

PROBLEM **11.4.** CLASSIFICATION OF ISOMETRIES

a. *Prove that on the plane, spheres, or hyperbolic planes, every isometry is the composition of one, two, or three reflections.*

Look back at what we did in the proofs of SAS and ASA. Also use Problem **11.3** which allows you to be more concrete when thinking about isometries because you only need to look at the effect of the isometries on any three non-collinear points that you pick. To get started on **11.4**, cut a triangle out of an index card and use it to draw two congruent triangles in different orientations on a sheet of paper. For example, see Figure 11.12. Now, can you move one triangle to the other by three (or fewer) reflections? You can use your cutout triangle for the intermediate steps.

Figure 11.12   Can you make triangles coincide with three reflections?

We have already showed in Problems **5.3** and **11.1** that the product of two reflections is a rotation (if the two reflection lines intersect) and a translation (if the two reflection lines have a common perpendicular, that is, if the two lines are parallel transports). Thus we can

b. *Prove that on the plane and spheres, every composition of two reflections is either the identity, a translation, or a rotation. What are the other possibilities on a hyperbolic plane? What happens if you switch the order of the two reflections?*

What are the different ways in which the two reflection lines can intersect? To fully answer this part on a hyperbolic plane you will have to use the result (that is proved in Problem **16.3**) that: *Two geodesics in a hyperbolic plane either intersect, have a common perpendicular, or are asymptotic.*

Your proof of Part **b** or Problem **11.1** probably already shows that

THEOREM 11.4. *On the plane, spheres, and hyperbolic planes, a rotation is determined by two intersecting reflection lines (geodesics), which lines depend only on the point of intersection and the angle between the lines.* See Figure 11.13.

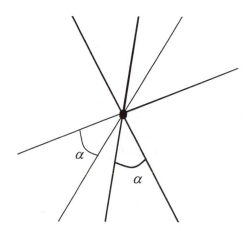

Figure 11.13   Two pairs of reflection lines, which determine the same rotation

c. *On the plane, spheres, and hyperbolic plane, the product of two rotations* (in general, about different points) *is a single rotation.*

Show how to geometrically determine the center and angle of the rotation.

Use Theorem 11.4 and write the composition of the two rotations with each rotation the composition of two reflections in a way that the middle two reflections cancel out. Be careful to keep track of the directions of the rotations.

    d. *On the plane, spheres, and hyperbolic planes, every composition of three reflections is either a reflection or a glide reflection. How can you tell which one?*

Approach using Theorem 11.4: Write the composition of three reflections, R_l, R_m, R_n, as either $(R_lR_m)R_n$ or $R_l(R_mR_n)$. If the reflection lines, l, m, n, intersect, then use Part **c** to replace the reflection lines within the parentheses with reflection lines that produce the same result. Try to produce the situation where the first (or last) reflection line is perpendicular to the other two. What is the situation if none of l, m, n intersect each other? This approach works well on the plane and spheres but is more difficult on a hyperbolic plane because the situation when all three lines do not intersect is more complicated.

Approach using 11.3 and triangles: Let $\triangle ABC$ be a triangle and let $\triangle A'B'C'$ be its image under the composition of the three reflections. Note that the two triangles cannot be directly congruent (see the discussion around Figure 6.5). Prolong two corresponding sides (say AB and $A'B'$) of the triangles to lines (geodesics) l and m. Then there are three cases: The lines l and m intersect, or have a common perpendicular, or are asymptotic.

    If Parts **a**, **b**, and **d** are true, then

    e. *Every isometry of the plane or a sphere or a hyperbolic plane is either a reflection, a translation, a rotation, a horolation, a glide reflection, or the identity.*

Notice that in our proofs of SAS and ASA and (probably) in your proof of Part **b** we only need to use reflections about the perpendicular bisectors or segments joining two points. Because, on a sphere, we only need great circles to join two points and for perpendicular bisectors, we can restate Part **b** on a sphere to read

> *Every isometry of a sphere is the composition of one, two, or three reflections through great circles.*

Thus, in particular, if there were a reflection of the sphere that was not through a great circle then that refection (being an isometry) would also be composition of one, two, or three reflections through great circles.

You now have powerful tools to make a classification of discrete strip patterns on the plane, spheres, and hyperbolic planes and finite patterns on the plane and hyperbolic planes.

PROBLEM 11.5   CLASSIFICATION OF DISCRETE STRIP PATTERNS

A strip pattern is *discrete* if every translation symmetry of the strip pattern is a multiple of some shortest translation.

    a. *Prove there are only seven strip patterns on the plane that are discrete.*

    b. *What are some non-discrete strip patterns?*

    c. *What happens with strip patterns on spheres and hyperbolic planes?*

Hint: Use Problem **11.4**.

PROBLEM 11.6   CLASSIFICATION OF FINITE PLANE PATTERNS

    a. *Show that any pattern on the plane with only finitely many symmetries has a center. That is, there is a point in the plane (not necessarily on the figure) such that every symmetry of the pattern leaves the point fixed. Is this true on spheres and hyperbolic planes?*

This was first proved by Leonardo da Vinci (1452–1519, Italian), and you can prove it too! Hint: Start by looking at what happens if there is a translation or glide symmetry.

> **b.** *Describe all the patterns on the plane and hyperbolic planes with only finitely many symmetries.*

Hint: Use Part **a**. What rotations are possible if there are only finitely many symmetries?

> **c.** *Describe all the patterns on the sphere that are finite and that have centers.*

Hint: If there is one center, then its antipodal point is necessarily also a center. *Why?*

If you take a cube with its vertices on a sphere and project from the center of the sphere the edges of the cube onto the sphere, then the result is a pattern on the sphere with only finitely many symmetries. This pattern does not fit with the plane patterns you found in Part **c** because this pattern has no center (on the sphere). See Problem **21.5** for more examples.

PROBLEM **11.7** REGULAR TILINGS WITH POLYGONS

We will now consider some special infinite patterns of the plane, spheres, and hyperbolic planes. These special patterns are called *regular tilings* (or *regular tessellations*, or *mosaics*).

> **DEFINITION.** *A **regular tiling** {n,k} of a geometric space (the plane, a sphere, or a hyperbolic plane) is made by taking identical copies of a regular n-gon (a polygon with n edges) and using these n-gons to cover every point in the space so that there are no overlaps except each edge of one n-gon is also the edge of another n-gon and each vertex is a vertex of k n-gons.*

You probably know the three familiar regular tilings of the plane: {3,6}, the regular tiling of the plane by triangles with six triangles coming together at a vertex; {4,4}, the regular tiling of the plane by squares with four squares coming together at a vertex; and {6,3}, the regular tiling of the plane by hexagons with three hexagons coming together at each vertex. There is only one way to tile a plane with regular hexagons; however, there are other ways to tile a plane with regular triangles and squares, but only one for each that is a *regular* tiling.

Note that each regular tiling can also be thought of as a (infinite) pattern. The notation {*n,k*} is called the **Schläfli symbol** of the tiling, named after Ludwig Schläfi (1814–1895, Swiss). We now study the possible regular tilings.

a. *Show that, {3,6}, {4,4}, {6,3} are the only Schläfli symbols that represent regular tilings of the plane.*

Focus on what happens at the vertices.

There are 14 more infinite (non-isomorphic) patterns on the plane, besides the three that come from regular tilings. These 17 patterns in the plane are often called **wall-paper patterns**. See Escher's (Maurits Cornelius Escher, 1898–1972, Dutch) drawings for examples and [**SG:** Budden] for proofs and more examples. For a complete exposition on patterns and tilings on the plane see [**SG:** Grünbaum].

b. *Find all the regular tilings of a sphere.*

Again, focus on what happens at the vertices. Remember that angles of a regular *n*-gon on the sphere are larger than the angles of the corresponding regular *n*-gon on the plane (*Why?*); and use Problems **7.1** and **7.4**.

There are more finite patterns of the sphere besides those given in Problem **11.6c** and **11.7b**. See a soccer ball for an example and [**SG:** Montesinos] for the complete classification.

c. *Show that each of the Schläfli symbols {n,k} with both n and k greater than 1 represents a regular tiling of the plane (Part **a**), or of a sphere (Part **b**), or of a hyperbolic plane.*

Again, focus on what happens at the vertices.

See [**SG:** Montesinos] for a discussion of other patterns and tilings on hyperbolic spaces.

*Geometric Meaning
of Abstract Group Terminology

The collection of symmetries of a geometric figure with the operation of composition is an **abstract group.** We showed above how the usual axioms of a group are satisfied. For example, the following figure has symmetry group $\{R_A, R_B, R_C, S_{1/3}, S_{2/3}, Id\}$ which is isomorphic as an abstract group to $\mathbf{D_3}$ the third dihedral group.

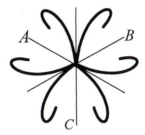

Figure 11.14   Isomorphic to D₃

If a figure has a subfigure, then the symmetry group of the subfigure is a **subgroup** of the symmetry group of the original figure. See Figure 11.15.

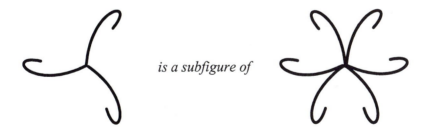

is a subfigure of

Figure 11.15   Subgroups

The symmetry group of the subfigure is $\{S_{1/3}, S_{2/3}, Id\}$, which is isomorphic as an abstract group to $\mathbf{Z_3}$, which is isomorphic to a subgroup of $\mathbf{D_3}$.

In $\mathbf{D_3}$ the two **cosets** of $\mathbf{Z_3}$, $Id\mathbf{Z_3}$ and $R_A\mathbf{Z_3} = R_B\mathbf{Z_3} = R_C\mathbf{Z_3}$, correspond to the two copies of the subfigure in Figure 11.16.

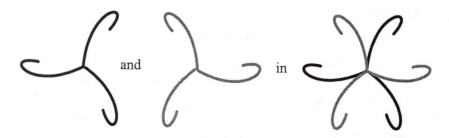

Figure 11.16   Cosets

In general, if G is the symmetry group of a figure and H is a subgroup that is the symmetry group of a subfigure in the figure, then the *cosets* of H correspond to the several congruent copies of the subfigure that exist within the larger figure.

Chapter 12

Dissection Theory

Oh, come with old Khayyám, and leave the Wise
To talk, one thing is certain, that Life flies;
One thing is certain, and the Rest is Lies;
The flower that once has blown for ever dies.
— Omar Khayyam, *Rubaiyat*[†]

What Is Dissection Theory?

Figure 12.1   Parallelogram

In showing that the parallelogram in Figure 12.1 has the same area as a rectangle with the same base and height (altitude), we can easily cut the parallelogram into two pieces and rearrange them to form the rectangle in Figure 12.2.

Figure 12.2   Equivalent by dissection to a rectangle

[†]From the translation by Edward Fitzgerald.

We say that two figures (F and G) are ***equivalent by dissection*** ($F =_d G$) if one can be cut up into a finite number of pieces and the pieces rearranged to form the other.

QUESTION: If two planar polygons have the same area, are they equivalent by dissection?

ANSWER: Yes! If the polygon is bounded on either the plane or on a sphere or on a hyperbolic plane.

You will prove these results about dissections in this chapter and the next and use them to look at the meaning of area. In this chapter you will show how to dissect any triangle or parallelogram into a rectangle with the same base. Then you will do analogous dissections on spheres and hyperbolic planes after first defining an appropriate analog of parallelograms and rectangles. After that you will show that two polygons on a sphere or on a hyperbolic plane that have the same area are equivalent by dissection to each other. The analogous result on the plane must wait until the next chapter.

The proofs and solutions to all the problems can be done using "$=_d$", but if you wish you can use the weaker notion of "$=_s$": We say that two figures (F and G) are ***equivalent by subtraction*** ($F =_s G$) if there are two other figures, S and S', such that $S =_d S'$ and $F \cup S =_d G \cup S'$, where F & S and G & S' intersect at most in their boundaries. Saying two figures are equivalent by subtraction means that they can be arrived at by removing equivalent parts from two initially equivalent figures, as in Figure 12.3.

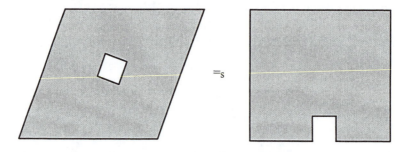

Figure 12.3   Equivalent by subtraction

If we cut out the two small squares as shown, we can see that the shaded portions of the rectangle and the parallelogram are equivalent by subtraction, but it is not at all obvious that one can be cut up and rearranged to form the other.

Equivalence by dissection is the stronger of the two methods of proof, and is generally preferable — it is much more obvious and in many ways, more convincing. All of the problems presented here can be proved by dissection, and we would urge you to use this method over equivalent by subtraction whenever you can.

Some of the dissection problems ahead are very simple, while some are rather difficult. If you think that a particular problem was so easy to solve that you may have missed something, chances are you hit the nail right on the head. For almost all of the problems, it is very helpful to make paper models and actually cut them up and fit the pieces together. Most of the dissection proofs will consist of two parts: First show where to make the necessary cuts, and then prove that your construction works, i.e., that all the pieces do in fact fit together as you say they do.

PROBLEM 12.1 DISSECT PLANE TRIANGLE AND PARALLELOGRAM

> **a.** *Show that on the plane every triangle is equivalent by dissection to a parallelogram with the same base no matter which base of the triangle you pick.*

Part **a** is fairly straightforward, so don't try anything complicated. You only have to prove it for the plane — a proof for spheres and hyperbolic planes will come in a later problem after we find out what to use in place of parallelograms. Make paper models, and make sure your method works for all possible triangles with any side taken as the base. In particular, make sure that your proof works for triangles whose heights are much longer than their bases. Also, you need to show that the resulting figure actually is a parallelogram.

> **b.** *Show that on a plane every parallelogram is equivalent by dissection to a rectangle with the same base and height.*

A partial proof of this was given in the introduction at the beginning of this chapter. But for this problem, your proof must also work for tall, skinny parallelograms, as shown in Figure 12.4, for which the given construction does not work. You may say that you can simply change the orientation of the parallelogram and use a long side as the base; but, as for Part **a**, we want a proof that will work no matter which side you choose as the base. Again, do not try anything too complicated, and you only have to work on the plane.

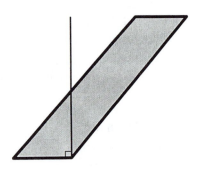

Figure 12.4   Tall, skinny parallelogram

DISSECTION THEORY ON SPHERES AND HYPERBOLIC PLANES

The above statements take on a different flavor when working on spheres and hyperbolic planes because one cannot construct parallelograms and rectangles, as such, on these spaces. We can define two types of polygons on spheres and hyperbolic spaces, and then restate the above two problems for these spaces. The two types of polygons are the Khayyam quadrilateral and the Khayyam parallelogram. These definitions were first put forth by the Persian geometer-poet Omar Khayyam (1048–1131) in the eleventh century AD [**AT:** Khayyam, 1931]. Through a bit of Western chauvinism, geometry books generally refer to these quadrilaterals as Saccheri quadrilaterals after the Italian priest and professor, Gerolamo Saccheri (1667–1733) who translated into Latin and extended the works of Khayyam and others.

A ***Khayyam quadrilateral (KQ)*** is a quadrilateral such that $AB \cong CD$ and $\angle BAD \cong \angle ADC \cong \pi/2$. A ***Khayyam parallelogram (KP)*** is a quadrilateral such that $AB \cong CD$ and AB is a parallel transport of DC along AD. In both cases, BC is called the ***base*** and the angles at its ends are called the ***base angles***.

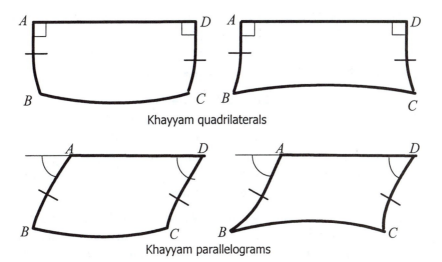

Figure 12.5   Khayyam quadrilaterals and parallelograms

PROBLEM 12.2   KHAYYAM QUADRILATERALS

a. *Prove that the base angles of a KQ are congruent.*

b. *Prove that the perpendicular bisector of the top of a KQ is also the perpendicular bisector of the base.*

c. *Show that the base angles are greater than a right angle on a sphere and less than a right angle on a hyperbolic plane.*

d. *A KQ on the plane is a rectangle and a KP on the plane is a parallelogram.*

To begin this problem, note that the definitions of KP and KQ make sense on the plane as well as on spheres and hyperbolic planes. The pictures in Figure 12.5 are deliberately drawn with a curved line for the base to emphasize the fact that the base angles are not necessarily congruent to the right angles. You should think of these quadrilaterals and parallelograms in terms of parallel transport instead of parallel lines. Everything you have learned about parallel transport and triangles on spheres and hyperbolic planes can be helpful for this problem. Symmetry can also be useful.

Now we are prepared to modify Problem **12.1** so that it will apply to spheres and hyperbolic planes.

PROBLEM 12.3 DISSECT SPHERICAL AND HYPERBOLIC TRIANGLES AND KHAYYAM PARALLELOGRAMS

    a. *Show that every hyperbolic triangle, and every small spherical triangle, is equivalent by dissection to a Khayyam parallelogram with the same base as the triangle.*

Try your proof from Problem **12.1**, as a first stab at the problem. You only need to look at a sphere. You should also look at the different proofs given for Problem **12.1**. The only difference between the plane and spheres and hyperbolic planes as far as this problem is concerned is that you must be more careful on spheres and hyperbolic planes because there are no parallel lines; there is only parallel transport. Some of the proofs for Problem **12.1** work well on a sphere or on a hyperbolic plane, and others do not. Remember that the base of a KP is the side opposite the given congruent angles.

    b. *Prove that every Khayyam parallelogram is equivalent by dissection to a Khayyam quadrilateral with the same base.*

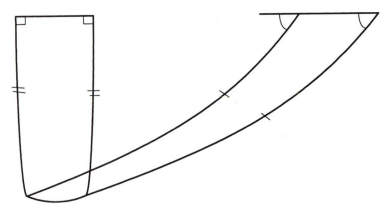

Figure 12.6   Dissecting KP's into KQ's

As with Part **a**, start with your planar proof and work from there. As before, your method must work for tall, skinny KPs. Once you have come up with a construction, you must then show that the pieces actually fit together as you say they do, and prove that the angles at the top are right angles. See Figure 12.6.

*PROBLEM **12.4** SPHERICAL POLYGONS DISSECT
TO LUNES

In the next chapter you will show that every polygon on the plane is equivalent by dissection to a square, and then we will use this and the Pythagorean Theorem to show that any two polygons with the same area are equivalent by dissection. This does not apply to spheres and hyperbolic planes because there are no squares on these surfaces. However, we have already shown in Problems **7.1** and **7.4** that two polygons (or triangles) on the same sphere have the same area if they have the same holonomy. Thus, every polygon on a sphere must have the same area as some lune with the same holonomy. Now we can show that not only do they have the same area but they are also equivalent by dissection.

> *Show that every simple* (sides intersect only at the vertices) *small polygon on a sphere is equivalent by dissection to a lune with the same holonomy.* That is, the angle of the lune is equal to
>
> $(½)(2\pi -$ sum of the exterior angles of the polygon).
>
> *Consequently, two simple small polygons on a sphere with the same area are equivalent by dissection.*

OUTLINE OF PROOF
The proof of this result can be completed by proving the following steps (or lemmas). (This proof was first suggested to me by my daughter, Becky, now Rebecca Wynne.)

1. *Every simple small polygon can be dissected into a finite number of small triangles, such that the holonomy of the polygon is the sum of the holonomies of the triangles.*

See Problem **7.5** but what is needed here is easier than **7.5**.

2. *Each small triangle is equivalent by dissection to a KQ with the same base and same holonomy.*

Check your solutions for Problems **12.2** and **12.3**.

3. *Two KQ's with the same base and the same holonomy (or base angles) are congruent.*

Match up the bases and see what you get.

4. *If two Δ's have the same base and the same holonomy, then they are equivalent by dissection.*

Put together the previous steps.

5. *Any Δ is equivalent by dissection to a lune with $\mathcal{H}(\Delta) = \mathcal{H}(\text{lune})$ = (twice the angle of the lune).*

Hint: A lune can also be considered as a triangle.

6. *Two simple small polygons on a sphere with the same area are equivalent by dissection to the same lune, and therefore are equivalent by dissection to each other.*

What is the union of two lunes?

The first four steps above will also work (with essentially the same proofs) on a hyperbolic plane. But there is no clear replacement for the biangles (which do not exist on a hyperbolic sphere). There is a proof of

THEOREM 12.4. *On a hyperbolic space, two simple* (the sides intersect only at the vertices) *polygons with the same area are equivalent by dissection.*

Two published proofs in English are in [**NE:** Millman & Parker], page 267, and [**Di:** Boltyanski], page 62. These proofs are similar, and both use the first four steps above and use the completeness of the real numbers (in the form of a version of the Intermediate Value Theorem). You can check that Becky's proof above does not use completeness. In addition, the proof of the same result on the plane (see the discussion between Problems **13.2** and **13.3**) also does not need the use of completeness axiom.

In the plane all the triangles with the same base and the same height have the same area and the vertices opposite the base of these triangles form a (straight) line not intersecting the line determined by the base. On a sphere, the situation is different. Your proof above should show that midpoints of the (non-base) edges lie on a great circle and the vertices opposite the base must lie on a curve equidistant from this great circle. See Figure 12.7.

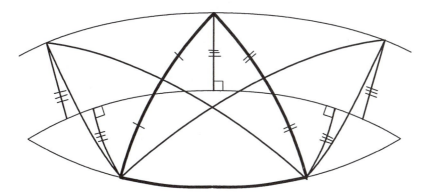

Figure 12.7   Triangles with same base and same area

Chapter 13

SQUARE ROOTS, PYTHAGORAS, AND SIMILAR TRIANGLES

> The diagonal of an oblong produces by itself both the areas
> which the two sides of the oblong produce separately.
> — Baudhayana, *Sulbasutram*, Sutra 48 [**AT: Baudhayana**]

In the last chapter, we showed that two polygons on a sphere with the same area are equivalent by dissection because both are equivalent to the same biangle. In this chapter, we will prove an analogous result on the plane by showing that every planar polygon is equivalent by dissection to a square. Before reading this chapter you should go through the introduction to dissection theory and Problem **12.1** at the beginning of Chapter 12.

In the process of exploring this dissection theory, we will follow a path through a corner of the forest of mathematics — a path that has delighted and surprised the author many times. We will bring with us the question: What are square roots? Along the way we will confront relationships between geometry and algebra of real numbers, in addition to similar triangles, the Pythagorean Theorem (the quote above is a statement of this theorem written before Pythagoras), and possibly the oldest written proof in geometry (at least 2,600 years old). This path will lead to the solutions of quadratic and cubic equations in Chapter 18. We will let the author's personal experience lead us on this path.

SQUARE ROOTS

When I was in eighth grade, I asked my teacher, "What is a square root?" I knew that the square root of N was a number whose square was equal to N, but where would I find it? (Hidden in that question is "How do I know it always exists?") I knew what the square roots of 4 and 9 were — no problem there. I even knew that the square root of 2 was the length of the diagonal of a unit square, but what of the square root of 2.5 or of π?

At first, the teacher showed me a square root table (a table of numerical square roots), but I soon discovered that if I took the number listed in the table as the square root of 2 and squared it, I got 1.999396, not 2. (Modern-day pocket calculators give rise to the same problem.) So I persisted in asking my question — What is the square root? Then the teacher answered by giving me THE ANSWER — the Square Root Algorithm. Do you remember the Square Root Algorithm — that procedure, similar to long division, by which it is possible to calculate the square root? Or perhaps more recently you were taught the "divide and average" method, a more efficient method that was known to Archytas of Tarentum (428–350 B.C., Greek) but today is often called *Newton's Method* after Isaac Newton (English, 1643–1727).

If A_1 is an approximation of the square root of N then the average of A_1 and N/A_1 is an even better approximation, which we could call A_2. And then the next approximation A_3 is the average of A_2 and N/A_2. In equation form this becomes

$$A_{n+1} = (1/2)(A_n + (N/A_n)).$$

For example, if $A_1 = 1.5$ is an approximation of the square root of 2, then

$$A_2 = 1.417\cdots, \quad A_3 = 1.414216\cdots$$

and so forth are better and better approximations.

But wait! Most of the time these algorithms do not calculate the square root — they only calculate approximations to the square root. The algorithms have an advantage over the tables because I could, at least in theory, calculate approximations as close as I wished. However, they are still only approximations and my question still remained — What is the square root these algorithms approximate?

My eighth-grade teacher then gave up, but later in college I found out that some modern mathematicians answer my question in the following way: "We make an assumption (the Completeness Axiom) which implies that the sequence of approximations from the Square Root

Algorithm must converge to some real number." And, when I continued to ask my question, I found that in modern mathematics the square root is a certain equivalence class of Cauchy (Augustin-Louis Cauchy, 1789–1857, French) sequences of rational numbers, or a certain Dedekind (Julius Dedekind, 1831–1916, German) cut. Finally, I let go of my question and forgot it in the turmoil of graduate school, writing my thesis and beginning my mathematical career.

Later, I started teaching the geometry course that is the basis for this book. One of the problems in the course is the following problem.

PROBLEM 13.1 A RECTANGLE DISSECTS INTO A SQUARE

Show that, on the plane, every rectangle is equivalent by dissection to a square.

SUGGESTIONS

Problems **13.1** and **13.2** can be done in any order. So if you get stuck on one problem, you can still go on to the other. In Problems **13.1** and **13.2**, it is especially important to make accurate models and constructions — rough drawings will not show the necessary length and angle relationships.

Problem **13.1** is one of the oldest problems in geometry, so you may have guessed (correctly!) that it is one of the more complex ones. This problem is interesting for more than just historical reasons. You are asked to prove that you can cut up any rectangle (into a finite number of pieces) and rearrange the pieces to form a square, as in Figure 13.1.

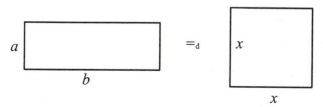

Figure 13.1   Dissecting a rectangle into a square

Because you are neither adding anything to the rectangle nor removing anything, the area must remain the same, so $ab = x^2$, or $x = \sqrt{ab}$. What you are really finding is a geometric interpretation of a square root. This was done already by Pythagoreans (6th Century B.C.) who called x the geometric mean of a and b.

Let us look at a proof that is similar to the proofs in many standard geometry textbooks:

Let $s = \sqrt{ab}$ be the side of the square equivalent by dissection to the rectangle with sides a and b. Place the square, $AEFH$, on the rectangle, $ABCD$, as shown in Figure 13.2. Draw ED to intersect BC in R and HF in K. Let BC intersect HF in G.[†] From the similar triangles $\triangle KDH$ and $\triangle EDA$ we have $HK/AE = HD/AD$, or

$$HK = (AE)(HD)/AD = s(a - s)/a = s - s^2/a = s - b.$$

Therefore, we have $\triangle EFK \cong \triangle RCD$, $\triangle EBR \cong \triangle KHD$.

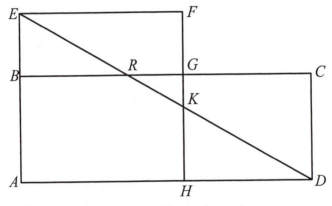

Figure 13.2   Textbook proof

I was satisfied with the proof until, in the second year of teaching the geometry course, I started sensing student uneasiness with it. As I listened to their comments, I noticed questions being asked by the students: "What is $\sqrt{ab}$?" "How do you find it?" Those used to be my questions!

The students and I also noticed that the facts used about similar triangles in the proof above are usually proved using the theory of areas of triangles. Thus, the proof could not be used as part of a concrete theory of areas of polygons, which was our purpose in studying dissection

[†] In the case that $ABCD$ is so long and skinny that K ends up between G and F, we can, by cutting $ABCD$ in half and stacking the halves, reduce the proof to the case above.

theory in the first place. Notice that the above proof also assumes the existence of the square, which in analysis is based on what is called the Completeness Axiom. The conclusion here seems to be that it would be desirable instead to construct the square root x. That started me on an exploration that continued on and off for many years.

Now let us solve a few problems in dissection theory.

Here are three methods for constructing x. For all three constructions we will use a rectangle like the one shown in Figure 13.1, with the longer side b as the base and the shorter side a as the height.

For both the first and second constructions, you need to know that a triangle with all vertices on the circumference of a circle and one side a diameter of the circle is a right triangle. See Figure 13.3. That the angle not on the diameter is a right angle follows from a more general result, which we will prove in Problem **14.1a**, but the special case we need is easy to see now. In particular, draw the line segment from the vertex to the center of the circle and note that two isosceles triangles are formed. By ITT, Problem **6.2**, there are congruent angles as marked; and, because the sum of the angles in a planar triangle is $180°$ (Problems **7.3b** or **10.2**) we can compute that

$$(2\alpha + \gamma) + (2\beta + \mu) = 360°$$

and, because $\gamma + \mu = 180°$, it follows immediately that $\alpha + \beta = 90°$.

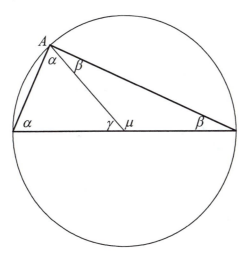

Figure 13.3 The angle at A is a right angle

For the first construction, Figure 13.4, take the rectangle and lay *a* out to the left of *b*. Use this base line as the diameter of a circle. The length *x* that you are looking for is the perpendicular line from the left side of the rectangle to where it intersects the circle.

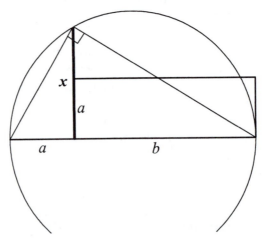

Figure 13.4   First construction of *x*

The second construction, Figure 13.5, is similar but this time put *a* on the inside of the base of the rectangle. Now the side *x* you are looking for is the segment from the lower left corner of the rectangle to the point at which a perpendicular rising from the place you put *a* intersects the circle.

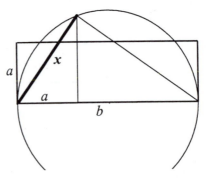

Figure 13.5   Second construction of *x*

The third construction, Figure 13.6, is a bit more algebraic than the others, and doesn't directly involve a circle.

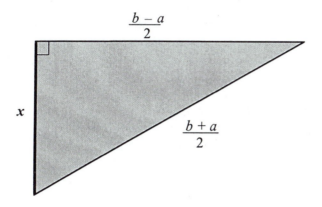

Figure 13.6   Third construction of x

This construction can be used together with the result of Problem **13.2** in order to obtain a proof by subtraction.

For all of these constructions, it is imperative that you use accurate models. Whatever method you choose, make the rectangle and the square overlap as much as possible, and see how to fit the other pieces in. Then, you have to prove that all of the sides and angles line up properly. Note that it is much better to solve this problem geometrically rather than by only trying to work out the algebra — actually do the construction and proceed from there. Finally, you don't have to use one of the constructions shown here. If these don't make sense to you, then find one of your own that does.

And remember not to use results about similar triangles because we normally need results about areas in order to prove these results as we will do in Problem **13.3**.

Pause, explore, and write out your ideas before reading further.

BAUDHAYANA'S SULBASUTRAM

While reading an unrelated article, I ran across an item that said that the problem of changing a rectangle into a square appeared in the *Sulbasutram*. "Sulbasutram" means "rules of the cord" and the several Sanskrit texts collectively called the *Sulbasutra* were written by the Vedic Hindus starting before 600 B.C. and are thought to be compilations of oral wisdom that may go back to 2000 B.C. (See, for example, A. Seidenberg *The Ritual Origin of Geometry* [**Hi**: Seidenberg].) These texts have prescriptions for building fire altars, or *Agni*. However, contained in the *Sulbasutra* are sections that constitute a geometry textbook detailing the geometry necessary for designing and constructing the altars. As far as I have been able to determine these are the oldest geometry (or even mathematics) textbooks in existence. There are at least four versions of the *Sulbasutram* by Baudhayana, Apastamba, Katyayana, and Manava. The geometric descriptions are very similar in these four books and I will only use Baudhayana's version here (see [**AT**: Baudhayana]).

The first chapter of Baudhayana's *Sulbasutram* contains geometric statements called "Sutra." Sutra 54 is what we asked you to prove in Problem **13.1**. It states

> If you wish to turn an oblong [here "oblong" means "rectangle"] into a square, take the shorter side of the oblong for the side of square. Divide the remainder into two parts and inverting join those two parts to two sides of the square. Fill the empty place by adding a piece. It has been taught how to deduct it.

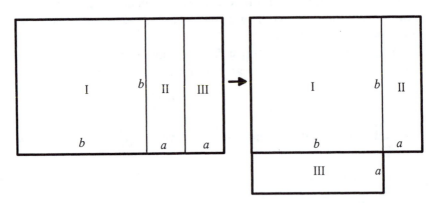

Figure 13.7   Sutra 54

So our rectangle has been changed into a figure with an "empty place," which can be filled "by adding a piece" (a small square). The result is a large square from which a small square has to be "deducted".

Now Sutra 51:

> If you wish to deduct one square from another square, cut off a piece from the larger square by making a mark on the ground with the side of the smaller square which you wish to deduct; draw one of the sides across the oblong so that it touches the other side; by this line which has been cut off the small square is deducted from the large one.

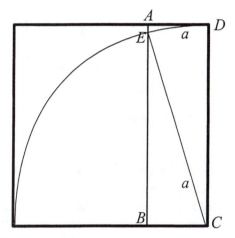

Figure 13.8   Sutra 51: Construction of side of square

We wish to deduct the small square (a^2) from the large square. Sutra 51 tells us to "scratch up" with the side of the smaller square — this produces the line AB and the oblong $ABCD$. Now, if we "draw" the side CD of the large square to produce an arc, then this arc intersects the other side at the point E. The sutra then claims that BE is side of the desired square whose area equals the area of the large square minus the area of the small square. This last assertion follows from Sutra 50, which we ask you to prove in Problem **13.2**.

See Figure 13.9 for the drawing that goes with Sutra 50:

> If you wish to combine two squares of different size into one, scratch up with the side of the smaller square a piece cut off from the larger one. The diagonal of this cutoff piece is the side of the combined squares.

Be sure you see why Sutra 50 is a statement of what we call the Pythagorean Theorem.

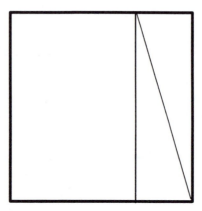

Figure 13.9   Sutra 50

A. Seidenberg, in an article entitled *The Ritual Origin of Geometry* [**Hi**: Seidenberg], gives a detailed discussion of the significance of Baudhayana's *Sulbasutram*. He argues that it was written before 600 B.C. (Pythagoras lived about 500 B.C. and Euclid about 300 B.C.). He gives evidence to support his claim that it contains codification of knowledge going "far back of 1700 B.C." and that knowledge of this kind was the common source of Indian, Egyptian, Babylonian, and Greek mathematics. Together Sutras 50, 51, and 54 describe a construction of a square with the same area as a given rectangle (oblong) and a proof (based on the Pythagorean Theorem) that this construction is correct. You can find stated in many books and articles that the ancient Hindus, in general, and the *Sulbasutram*, in particular, did not have proofs or demonstrations, or they are dismissed as being "rare." However, there are several sutras in [**AT**: Baudhayana] similar to the ones discussed above. I suggest you decide for yourself to what extent they constitute proofs or demonstrations.

Baudhayana avoids the Completeness Axiom by giving an explicit construction of the side of the square. The construction can be summarized in Figure 13.10.

This is the same as Euclid's construction in Proposition II-14 (see [**AT**: Euclid, *Elements*], page 409). But Euclid's proof is much more complicated. Note that neither Baudhayana nor Euclid gives a proof of Problem **13.1** because the use of the Pythagorean Theorem obscures the dissection. However, they do give a concrete construction and a proof that the construction works.

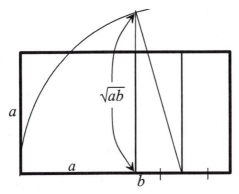

Figure 13.10   Baudhayana's construction of square root

Both Baudhayana and Euclid prove the following theorem, which uses equivalence by subtraction.

> **THEOREM.** *For every rectangle R there are squares S_1 and S_2 such that $R + S_2$ is equivalent by dissection to $S_1 + S_2$ and thus R and S_1 have the same area.*

Notice that both Baudhayana's and Euclid's proofs of this theorem and your proof of Problem **13.1** avoid assuming that the square root exists (and, thus, avoid the Completeness Axiom). They also avoid using any facts about similar triangles. These proofs explicitly construct the square and show in an elementary way that its area is the same as the area of the rectangle. There is no need for the area or the sides of the rectangle to be expressed in numbers. Also given a real number, *b*, the square root of *b* can be constructed by using a rectangle with sides *b* and 1. In Problem **13.3** we will use these techniques to prove basic properties about similar triangles.

So, finally, we have an answer to our question — What is a square root? It is *"an* answer" because many other solutions given from different points of view may be on the horizon. In particular, the *Appendix* to this book contains a geometric method based on Baudhayana that finds arbitrary accurate numerical approximations to many square roots and does so in a way that is computationally more efficient than the "divide and average" method (*Archytas/Newton's Method*) that is described at the beginning of this chapter.

PROBLEM 13.2   EQUIVALENCE OF SQUARES

Prove the following: On the plane, the union of two squares is equivalent by dissection to another square.

SUGGESTIONS

This is closely related to the Pythagorean Theorem. There are two general ways to approach this problem: You can use Problem **13.1** or you can prove it on its own, which will result in a proof of the Pythagorean Theorem — hence, you can't use the Pythagorean Theorem to solve this problem because you will be proving it!

To see how this problem relates to the Pythagorean Theorem, think about the following statement of the Pythagorean Theorem:

The square on the hypotenuse is equal to the sum of the squares on the other two sides.

This is not just an algebraic equation — **the squares referred to are actual geometric squares**.

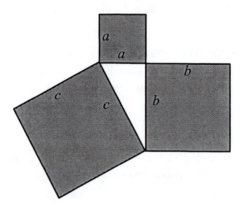

Figure 13.11   Pythagorean Theorem

As in the other dissection problems, make the three squares coincide as much as possible, and see how you might get the remaining pieces to overlap. You might start by reflecting the square with side c over the side c on the triangle. Then prove that the construction works as you say it does.

ANY POLYGON CAN BE DISSECTED INTO A SQUARE

If you put together Problems **12.1**, **13.1**, and **13.2**, a surprising result is created.

> **THEOREM.** *On the plane, every polygon is equivalent by dissection to a square.*

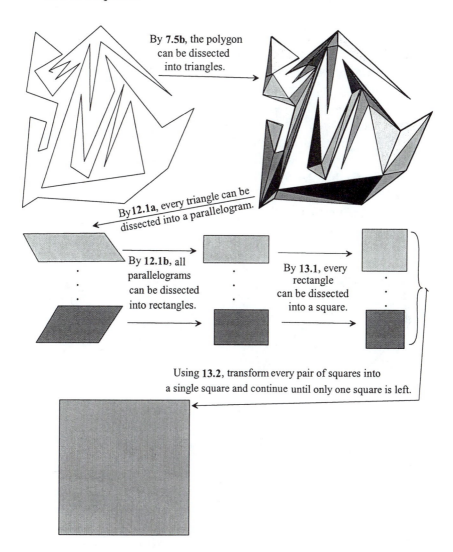

Figure 13.12   Every polygon dissects into a square

PROBLEM **13.3** SIMILAR TRIANGLES

Near the beginning of this chapter we gave a textbook proof of Problem
13.1, which used properties of similar triangles. Later you found a proof
that did not need to use similar triangles. Remember that in our discussion of AAA we said that two triangles were *similar* if their corresponding angles are congruent. Now you are ready to give a dissection proof
of

   a. **AAA criterion:** *If two triangles are similar, then the corre-
        sponding sides of the triangles are in the same proportion to one
        another.*

SUGGESTIONS

Look at your proof of Problem **13.1**. It probably shows implicitly that
Problem **13.3** holds for a pair of similar triangles in your construction.
For more generality, let θ be one of the angles of the triangles and place
the two θ's in VAT position in such a way that they form two parallelograms as in Figure 13.13.

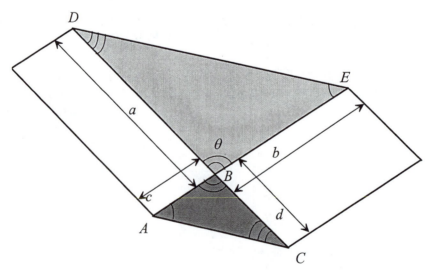

Figure 13.13   Similar triangles

Show that the two parallelograms are equivalent by dissection and use that result to show that $ac = bd$ or in equivalent form $a/d = c/b$. You may find it clearer if you start by looking at the special case of $\theta = \pi/2$.

b. SAS criterion: *If two triangles have an angle in common and if the corresponding sides of the angle are in the same proportion to each other, then the triangles are similar.*

Draw the two triangles with their common angle coinciding. Then show that the opposite sides are parallel. You will probably have to use Part **a**.

THREE-DIMENSIONAL DISSECTIONS AND HILBERT'S THIRD PROBLEM

In 1900, David Hilbert delivered a lecture before the International Congress of Mathematicians in which he listed ten problems "from the discussion of which an advancement of science may be expected." In a later paper he expanded these to 23 problems, which are now called *Hilbert's Problems*.

Hilbert's Third Problem. *Is it possible to specify two tetrahedra of equal bases and equal altitudes which can in no way be split up into congruent tetrahedra, and which cannot be combined with congruent tetrahedra to form two polyhedra which themselves could not be split up into congruent tetrahedra?* See [**Di:** Sah] and [**Di:** Boltyanski] for a discussion of this problem and its history.

Shortly after Hilbert's lecture, Max Dehn (1878–1952, Germany, USA) found such tetrahedra and also proved that a regular tetrahedron is not equivalent by dissection to a cube. Thus there is no possibility of dissecting polyhedra into cubes. To show these results, Dehn proved the following:

THEOREM. *If P and Q are two polyhedra in 3-space that are equivalent by dissection (or by subtraction), then the dihedral angles (see Chapter 21) of P are, mod π, a rational linear combination of the dihedral angles of Q. That is, if α_i are the* *dihedral angles of P and β_j are the dihedral angles of Q, then there are integers n_i, m_j, a, b such that*

$$\sum_i n_i\alpha_i + a\pi = \sum_j m_j\beta_j + b\pi.$$

Chapter 14

CIRCLES IN THE PLANE

Q: How does a geometer capture a lion in the desert?
A: Build a circular cage in the desert, enter it, and lock
it. Now perform an inversion with respect to the cage. Then
you are outside and the lion is locked in the cage.
— a mathematical joke from before 1938

We now study some properties of planar circles that do not have analogous results on a sphere (or hyperbolic space). We will use these results
about circles in the plane only for the detailed study of the upper half
plane model of the hyperbolic plane. These results are also often studied
for their own interest because this method can significantly simplify
solving some geometrical problems.

To study Chapter 14, the only result needed from Chapters 10–13 is

PROBLEM 13.3: The **AAA criterion** and **SAS criterion** for
similar triangles.

If you are willing to assume these criteria for similar triangles, then you
can work through Chapter 14 without Chapters 10–13.

PROBLEM 14.1 ANGLES AND POWER POINTS OF PLANE CIRCLES

a. *If an arc of a circle subtends an angle 2α from the center of the
circle, then the same arc subtends an angle α from any point on
the circumference.*

Use Figures 14.1 and 14.2. Draw a segment from the center of the circle to the point *A* and use ITT. Note the four different locations for *A*.

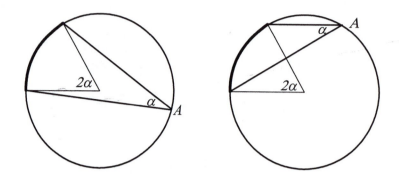

Figure 14.1   Angles subtended from outside the arc

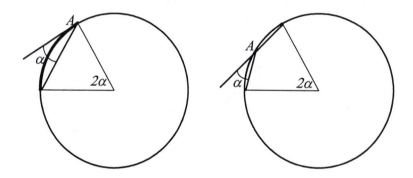

Figure 14.2   Angles subtended from on the arc

b. *If two lines through a point P intersect a circle at points A, A' (possibly coincident) and B, B' (possibly coincident), then*

$$|PA| \times |PA'| = |PB| \times |PB'|$$

*This product is called the **power of the point P with respect to the circle**.*

Use Figures 14.3 and 14.4 and draw the segment joining *A* to *B'* and the segment joining *A'* to *B*. Then apply Part **a** and look for similar triangles.

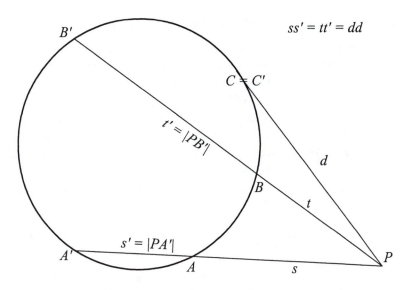

Figure 14.3   Power of a point outside with respect to a circle

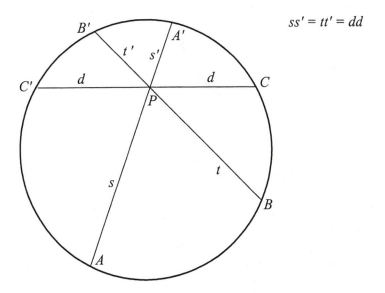

Figure 14.4   Power of a point inside with respect to a circle

PROBLEM 14.2   INVERSIONS IN CIRCLES

DEFINITIONS. An *inversion with respect to a circle* Γ is a transformation from the extended plane (the plane with ∞, the point at infinity, added) to itself that takes *C*, the center of the circle, to ∞ and vice versa, and that takes a point at a distance *s* from the center to the point on the same ray (from the center) that is at a distance of r^2/s from the center, where *r* is the radius of the circle. See Figure 14.5. We call (*P*,*P'*) an *inversive pair* because (as the reader can check) they are taken to each other by the inversion. The circle Γ is called the *circle of inversion*.

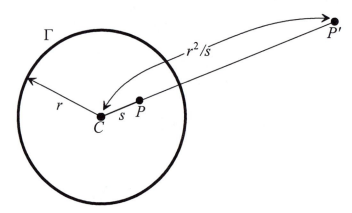

Figure 14.5   Inversion with respect to a circle

Note that an inversion takes the inside of the circle to the outside and vice versa and that the inversion takes any line through the center to itself. Because of this, inversion can be thought of as a reflection in the circle. See also Problem **14.3c** for the close connection between inversions in the plane and reflections on a sphere.

We strongly suggest that the reader play with inversions by using dynamic geometry software such as *Geometers Sketchpad®*, *Cabri®*, or *Cinderella®*. Using any of these you may construct the image, *P'*, of *P* under the inversion through the circle Γ as follows:

> *If P is inside* Γ *then draw through P the line perpendicular to the ray CP. Let S and R be the intersections of this line with* Γ*. Then P' is the intersection of the lines tangent to* Γ *at S and R.*

(To construct the tangents, note that lines tangent to a circle are perpendicular to the radius of the circle.)

If P is outside Γ then draw the two tangent lines from P to Γ. Let S and R be the points of tangency on Γ. Then P′ is the intersection of the line SR with the ray CP. (The points S and R are the intersections of Γ with the circle with diameter CP. Why?)

Sample constructions can be found at the *Experiencing Geometry* website: **http://www.math.cornell.edu/~dwh/books/eg2000** .

    a. *Prove that these constructions do construct inversive pairs.*

The purpose of this part is to explore and better understand inversion, but it will not be directly used in the other parts of this problem.

    b. *Show that an inversion takes each circle orthogonal to the circle of inversion to itself.* See Figure 14.6.

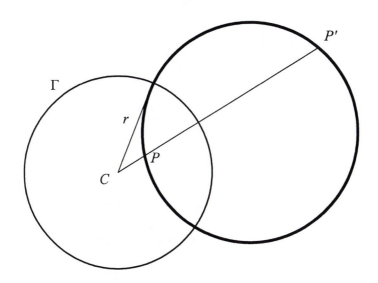

Figure 14.6   A circle orthogonal to Γ inverts to itself

Two **circles are orthogonal** if, at each point of intersection, the angle between the tangent lines is 90°. (Note that, at these points, the radius of one circle is tangent to the other circle.)

  c. *Show that an inversion takes a circle through the center of inversion to a line not through the center, and vice versa. What happens in the special cases when either the circle or the straight line intersects the circle of inversion?* Note that the line is parallel to the line tangent to the circle at *C*.

Look at Figure 14.7 and prove that $\triangle CPQ$ and $\triangle CQ'P'$ are similar triangles.

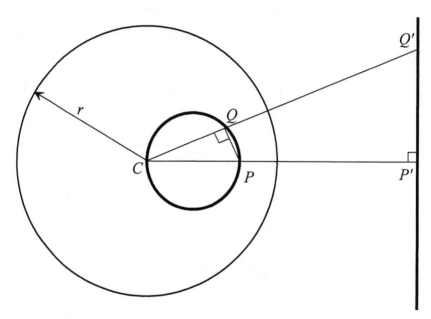

Figure 14.7   Circles through the center invert to lines

  d. *An inversion takes circles not through the center of inversion to circles not through the center.* Note: The circumference of a circle inverts to another circle but the centers of these circles are on the same ray from *C* though **not** an inversive pair.

Look at Figure 14.8. If *P*, *Q*, *X* invert to *P'*, *Q'*, *X'*, then show that

$$\angle P' X' Q' = \angle P X Q = \text{right angle},$$

by looking for similar triangles. Thus, argue that as X varies around the circle with diameter PQ then X' varies around the circle with diameter $Q'P'$.

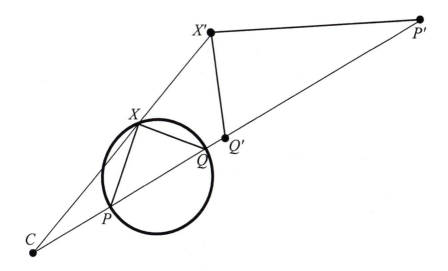

Figure 14.8   Circles invert to circles

e. *Inversions are conformal.* A transformation is called ***conformal*** if it preserves the measure of every angle.

Look at two lines that intersect and form an angle at P. Look at the images of these lines.

*PROBLEM **14.3** APPLICATIONS OF INVERSIONS

a. *Show that for the linkage in Figure* 14.9 *the points P and Q are the inversions of each other through the circle of inversion with center at C and radius* $r = \sqrt{s^2 - d^2}$.

Draw the circle with center R and radius d and note that C, P, Q, are collinear.

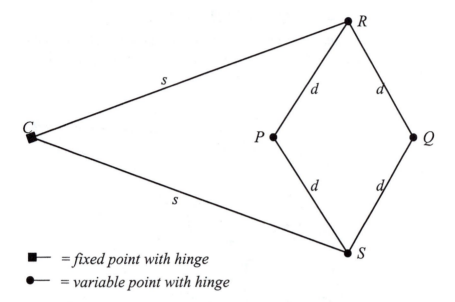

■— = *fixed point with hinge*
●— = *variable point with hinge*

Figure 14.9   A linkage for constructing an inversion

b. *Show that the point Q in the linkage in Figure 14.10 always traces a straight line.*

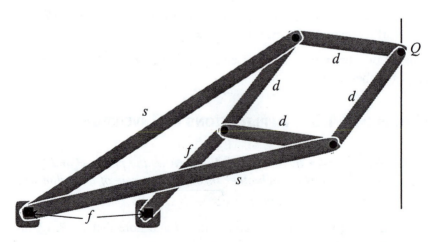

Figure 14.10   Linkage for drawing a straight line

The following result demonstrates the close connection between inversion through a circle in the plane and reflections through great circles on a sphere.

c. *Let Σ be a sphere tangent at its South Pole to the plane Π and let f: Σ → Π be a stereographic projection from the North Pole. If Γ is the circle that is the image under f of the equator and if g is the intrinsic (or extrinsic) reflection of the sphere through its equator (or equatorial plane), then show that the transformation f∘g∘f⁻¹ is the inversion of the plane with respect to the circle Γ.*

Imagine a sphere tangent to a plane at its South Pole, *S*. **Stereographic projection** is the transformation that maps each point, *Q*, (other than the North Pole, *N*) on the sphere to the point on the plane that is on the ray from *N* to *Q*. See Problem **15.4**. Stereographic projection was known already to Hipparchus (Greek, Second Century B.C.). Show that the triangle Δ*SNP* is similar to Δ*RNS*, which is congruent to Δ*QSN*, which is similar to Δ*SP'N*.

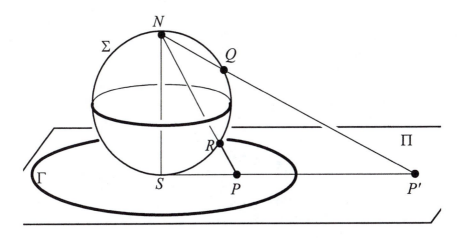

Figure 14.11   Stereographic projection and inversion

Part **c** provides another route to prove that inversions are conformal. In Problem **15.4** we will show that *f*, stereographic projection, is conformal. In addition, *g* (being an isometry) is conformal. Thus, the inversion *f*∘*g*∘*f*⁻¹ is conformal.

Chapter 15

PROJECTIONS OF A SPHERE ONTO A PLANE

Geography is a representation in picture of the whole known world together with the phenomena which are contained therein.

... The task of Geography is to survey the whole in its proportions, as one would the entire head. For as in an entire painting we must first put in the larger features, and afterward those detailed features which portraits and pictures may require, giving them proportion in relation to one another so that their correct measure apart can be seen by examining them, to note whether they form the whole or a part of the picture. ... Geography looks at the position rather than the quality, noting the relation of distances everywhere, ...

It is the great and the exquisite accomplishment of mathematics to show all these things to the human intelligence ...
— Claudius Ptolemy, *Geographia*, Book One, Chapter I

A major problem for map makers (cartographers) since Ptolemy (approx. 85–165 A.D., Alexandria, Egypt) and before is how to represent accurately a portion of the surface of a sphere on the plane. It is the same problem we have been having when making drawings to accompany our discussions of the geometry of the sphere. We shall use the terminology used by cartographers and differential geometers to call any one-to-one function from a portion of a sphere onto a portion of a plane a *chart*. As Ptolemy states in the quote above, we would like to represent the sphere on the plane so that proportions (and thus angles) are preserved and the relative distances are accurate. In this chapter we will study various charts for spheres. We will need properties of similar triangles that are investigated in Problem **13.3a**.

PROBLEM 15.1   CHARTS MUST DISTORT

It is impossible to make a chart without some distortions.

> *Which results that you have studied so far show that there must be distortions when attempting to represent a portion of a sphere on the plane?*

For a history and mathematical descriptions of charts of the sphere, see [**Hi:** Snyder].

Nevertheless, there are projections (charts) from a part of a sphere to the plane that do take geodesic segments to straight lines, that is, that preserve the shape of straight lines. There are other projections that preserve all areas. There are still other projections that preserve the measure of all angles. In this chapter, we will study these three types of projections on a sphere, and in Chapter 16 we will look at projections of hyperbolic planes.

PROBLEM 15.2   GNOMIC PROJECTION

Imagine a sphere resting on a horizontal plane. See Figure 15.1. A *gnomic projection* is obtained by projecting from the center of a sphere onto the plane. Note that only the lower open hemisphere is projected onto the plane; that is, if X is a point in the lower open hemisphere, then its gnomic projection is the point, $g(X)$, where the ray from the center through X intersects the plane.

a. *Show that a gnomic projection takes the portions of great circles in the lower hemisphere onto straight lines in the plane. (A mapping that takes geodesic segments to geodesic segments is called a **geodesic mapping**.)*

b. *Gnomic projection is often used to make navigational charts for airplanes and ships. Why would this be appropriate?*

Hint: Start with our extrinsic definition of great circle.

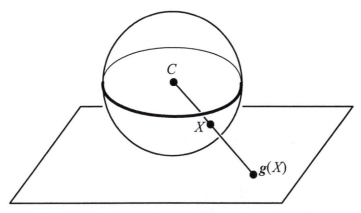

Figure 15.1   Gnomic projection

PROBLEM **15.3**   CYLINDRICAL PROJECTION

Imagine a sphere of radius r, but this time center it in a vertical cylinder of radius r and height $2r$. The *cylindrical projection* is obtained by projecting from the axis of the cylinder, which is also a diameter of the sphere; that is, if X is a point (not the North or South Poles) on the sphere and $O(X)$ is the point on the axis at the same height as X, then X is projected onto the intersection of the cylinder with the ray from $O(X)$ to X. See Figure 15.2.

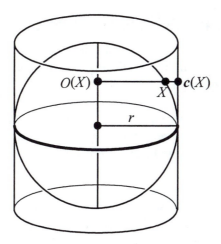

Figure 15.2   Cylindrical projection

    a. *Show that cylindrical projection preserves areas.* (Mappings that preserve area are variously called ***area-preserving*** or ***equiareal***.)

    Geometric Approach: Look at an infinitesimal piece of area on the sphere bounded by longitudes and latitudes. Check that when it is projected onto the cylinder the horizontal dimension becomes longer but the vertical dimension becomes shorter. Do these compensate for each other?

    *Analytic Approach: Find a function f from a rectangle in the (z, θ)-plane onto the sphere and a function h from the same rectangle onto the cylinder such that $c(f(z, \theta)) = h(z, \theta)$. Then use the techniques of finding surface area from vector analysis. (For two vectors A, B, the magnitude of the cross product $|A \times B|$ is the area of the parallelogram spanned by A and B. An element of surface area on the sphere can be represented by $|f_z \times f_\theta| \, dz \, d\theta$, the cross product of the partial derivatives.)

    We can easily flatten the cylinder onto a plane and find its area to be $4\pi r^2$. We thus conclude

    b. *The (surface) area of a sphere of radius r is $4\pi r^2$.*

PROBLEM 15.4   STEREOGRAPHIC PROJECTION

Imagine the same sphere and plane, only this time project from the uppermost point (North Pole) of the sphere onto the plane. This is called ***stereographic projection***.

    a. *Show that stereographic projection preserves the sizes of angles.* (Mappings that preserve angles are variously called ***angle-preserving***, ***isogonal***, or ***conformal***.)

SUGGESTIONS

There are several approaches for exploring this problem. Using a purely geometric approach requires visualization but only very basic geometry. An analytic approach requires knowledge of the differential of a function from $\mathbf{R}^2$ into $\mathbf{R}^3$.

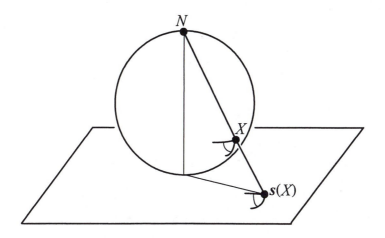

Figure 15.3   Stereographic projection is angle-preserving

Geometric Approach: An angle at a point X on the sphere is determined by two great circles intersecting at X. Look at the two planes that are determined by the North Pole N and vectors tangent to the great circles at X. Notice that the intersection of these two planes with the horizontal image plane determines the image of the angle. Because the 3-dimensional figure is difficult for many of us to imagine in full detail, you may find it helpful to consider what is contained in various 2-dimensional planes. In particular, consider the plane determined by X and the North and South Poles, the plane tangent to the sphere at X, and the planes tangent to the sphere at the North and South Poles. Determine the relationships among these planes.

*Analytic Approach: Introduce a coordinate system and find a formula for the function s^{-1} from the plane to the sphere, which is the inverse of the stereographic projection s. Use the differential of s^{-1} to examine the effect of s^{-1} on angles. You will need to use the dot (inner) product and the fact that the differential of s^{-1} is a linear transformation from the (tangent) vectors at $s(X)$ to the tangent vectors at X.

   b. *Show that stereographic projection takes circles to circles.* (Such mappings are called ***circle-preserving***.)

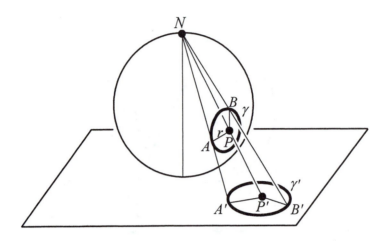

Figure 15.4   Stereographic projection is circle-preserving

SUGGESTIONS

Let γ be a circle on the sphere with points A and B and let γ', A', B' be their images under stereographic projection. Form the cone that is tangent to the sphere along the circle γ and let P be its cone point (note that P is not on the sphere). See Figure 15.4. Thus the segments BP and AP are tangent to the sphere and have the same length r. Look in the plane determined by N, A, and P and show that $\angle APP'$ is congruent to $\angle A'P'P$. You probably have already proved this in Part **a**; if not, look at the intersection of the plane determined by N, A, and P and the plane tangent to the North Pole N. Then use similar triangles to show that

$$|A'P'| = r\left(\frac{|NP'|}{|NP|}\right),$$

and thus γ' is a circle with center at P'.

Chapter 16

PROJECTIONS (MODELS) OF HYPERBOLIC PLANES

> In recent times the mathematical public has begun to occupy itself with some new concepts which seem to be destined, in the case they prevail, to profoundly change the entire order of classical geometry.
>
> — E. Beltrami (1868), when he developed the projective disk model (Problem **16.5**)

In this chapter we will study projections of a hyperbolic plane onto the plane and use these models to prove some results about the geometry of hyperbolic planes. In the case of hyperbolic planes it is customary to call these "*models*" instead of "projections" because it was thought that there were no surfaces that were hyperbolic planes. As in the case of spherical projections, any models must distort geometric properties of hyperbolic planes, and with them it is more difficult to gain the intrinsic and intuitive experiences that are possible with the hyperbolic surfaces discussed in Chapter 5. Nevertheless, these models do give the most analytically accurate picture of hyperbolic planes and allow for more accurate and precise constructions and proofs. In order to study these models and to connect them to hyperbolic surfaces we will need the results on circles from Chapter 14 and an analytic sophistication that is not necessary in other chapters in this book. However, no technical results from analysis are needed. The reader may bypass most of the analytic technicalities (which occur in Problems **16.1** and **16.2**) if the reader is willing to assume the results of Problem **16.2**, which make the connections between an annular hyperbolic plane and the upper half plane model and prove which curves in the upper half plane correspond to geodesics in the annular hyperbolic plane. The basic properties of geodesics and

193

constructions in the upper half plane model (and therefore in annular hyperbolic planes) are investigated in Problem **16.3**. We further our work on the area of triangles by investigating in Problem **16.4** ideal and 2/3-ideal triangles. Other popular models of hyperbolic planes are contained in **16.5** (Poincaré disk model) and **16.6** (projective disk model).

*PROBLEM 16.1   THE UPPER HALF PLANE MODEL

First, we will define coordinates on the annular hyperbolic plane that will help us to study it. The reader should review the description of the annular hyperbolic plane in Chapter 5. Let ρ be the fixed inner radius of the annuli and let H_δ be the approximation of the annular hyperbolic plane constructed, as in Chapter 5, from annuli of radius ρ and thickness δ. On H_δ pick the inner curve of any annulus and, calling it the *base curve*, pick a positive direction on this curve and pick any point on this curve and call it the origin O. We can now construct an (intrinsic) coordinate system x_δ: $R^2 \rightarrow H_\delta$ by defining $x_\delta(0,0) = O$, $x_\delta(w,0)$ to be the point on the base curve at a distance w from O, and $x_\delta(w,s)$ to be the point at a distance s from $x_\delta(w,0)$ along the radial geodesic through $x_\delta(w,0)$, where the positive direction is chosen to be in the direction from outer- to inner-curve of each annulus. Such coordinates are often called **geodesic rectangular coordinates**. See Figure 16.1.

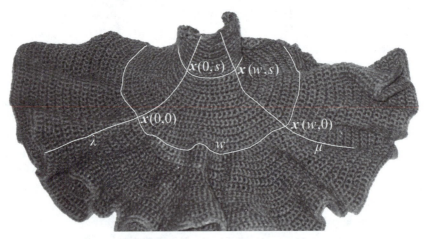

Figure 16.1   Geodesic rectangular coordinates on annular hyperbolic plane

a.   *Show that the coordinate map **x** is one-to-one and onto from the whole of R^2 onto the whole of the annular hyperbolic plane. What maps to the annular strips, and what maps to the radial geodesics?*

b.   *Let λ and μ be two of the radial geodesics described in Part **a**. If the distance between λ and μ along the base curve is w, then show that the distance between them at a distance $s = n\delta$ from the base curve is, on the paper hyperbolic model,*

$$w\left(\frac{\rho}{\rho+\delta}\right)^n = w\left(\frac{\rho}{\rho+\delta}\right)^{s/\delta}.$$

Now take the limit as $\delta \to 0$ to show that the distance between λ and μ on the annular hyperbolic plane is

$$w \exp(-s/\rho).$$

Thus, the coordinate chart **x** preserves (does not distort) distances along the (vertical) second coordinate curves but at $x(a,b)$ the distances along the first coordinate curve are distorted by the factor of $\exp(-b/\rho)$ when compared to the distances in R^2. To be more precise:

DEFINITION. Let **y**: $A \to B$ be a map from one metric space to another, and let $t \mapsto \lambda(t)$ be a curve in A. Then, the ***distortion*** of **y** along λ at the point $p = \lambda(0)$ is defined as

$$\lim_{x \to 0} \frac{\text{the arc length along } y(\lambda) \text{ from } y(\lambda(x)) \text{ to } y(\lambda(0))}{\text{the arc length along } \lambda \text{ from } \lambda(x) \text{ to } \lambda(0)}.$$

In the case of the above coordinate curves, λ is the path in R^2, $t \mapsto (a+t, b)$ or $t \mapsto (a, b+t)$, and the distortions of **x** along the coordinate curves are

$$\lim_{x \to 0} \frac{\text{the arc length from } x(a+t, b) \text{ to } x(a, b)}{|(a+t, b) - (a, b)|} =$$

$$= \frac{t\exp(-b/\rho)}{t} = \exp(-b/\rho)$$

and

$$\lim_{x \to 0} \frac{\text{the arc length from } x(a, b+t) \text{ to } x(a,b)}{|(a, b+t) - (a, b)|} = \frac{t}{t} = 1.$$

We seek a change of coordinates that will distort distances equally in both directions. The reason for seeking this change is that (as we will see below) if distances are distorted the same in both coordinate directions then the chart will preserves angles. (Remember, we call such a chart *conformal*.)

We cannot hope to have no distortion in both coordinate directions (if there were no distortion then the chart would be an isometry), so we try to make the distortion in the second coordinate direction the same as the distortion in the first coordinate direction. After a little experimentation we find that the desired change is

$$z(x,y) = x(x, \rho \ln(y/\rho)),$$

with the domain of z being the upper half plane

$$R^{2+} \equiv \{ (x,y) \in R^2 \mid y > 0 \},$$

where x is the geodesic rectangular coordinates defined above. This is usually called the **upper half plane model** of the hyperbolic plane. The upper half plane model is a convenient way to study the hyperbolic plane — think of it as a map of the hyperbolic plane in the same way that we use planar maps of the spherical surface of the earth.

c. *Show that the distortion of z along both coordinate curves*

$$x \to z(x,b) \quad and \quad y \to z(a,y)$$

at the point $z(a,b)$ is ρ/b.

It may be best to first try this for $\rho = 1$. For the first coordinate direction, use the result of Part **c**. For the second coordinate direction, use the fact that the second coordinate curves in geodesic rectangular coordinates are parametrized by arc length. Use first-semester calculus where necessary.

Lemma. *If the distortion of z at the point $p = (a,b)$ is the same (say $\Delta(p)$) along each coordinate curve, then the distortion of z at (a,b) has the same value along any other curve $\lambda(t) =$*

z(x(t),y(t)) that passes through p; and z preserves angles at p (that is, z is conformal).

Proof. Suppose that $\lambda(0) = (x(0),y(0)) = (a,b) = p$. Assuming that the annular hyperbolic plane can be locally isometrically (that is, preserving distances and angles) imbedded in 3-space (see Part **d**), the distortion of z along λ at p is

$$\lim_{t\to0} \frac{\text{the arc length along } z(\lambda) \text{ from } z(\lambda(t)) \text{ to } z(\lambda(0))}{\text{the arc length along } \lambda \text{ from } \lambda(t) \text{ to } \lambda(0)} =$$

$$= \lim_{t\to0} \frac{\frac{1}{|t|}[\text{the arc length along } z(\lambda) \text{ from } z(\lambda(x)) \text{ to } z(\lambda(0))]}{\frac{1}{|t|}[\text{the arc length along } \lambda \text{ from } \lambda(x) \text{ to } \lambda(0)]} =$$

$$= \frac{\text{speed of } z(\lambda(t)) \text{ at } t = 0}{\text{speed of } \lambda(t) \text{ at } t = 0} = \frac{\left|\frac{d}{dt}z(\lambda(t))\right|_{t=0}}{\left|\frac{d}{dt}\lambda(t)\right|_{t=0}} = \frac{|(z\circ\lambda)'(0)|}{|\lambda'(0)|}.$$

In particular, along the first coordinate curve $t \mapsto (a+t,b)$, the distortion is

$$\frac{\left|\frac{d}{dt}z(a+t,b)\right|_{t=0}}{\left|\frac{d}{dt}(a+t,b)\right|_{t=0}} = \text{the norm of the partial derivative, } |z_1(p)|.$$

Similarly, the distortion along the second coordinate curve is $|z_2(p)|$. The velocity vector of the curve $z(\lambda(t)) = z(x(t),y(t))$ at p is

$$\tfrac{d}{dt}z(x(t),y(t))_{t=0} = z_1(p)\tfrac{d}{dt}x(t)_{t=0} + z_2(p)\tfrac{d}{dt}y(t)_{t=0}.$$

Thus, the velocity vector, $(z\circ\lambda)'$ at $t = 0$ is a linear combination of the partial derivative vectors, $z_1(p)$ and $z_2(p)$ and, thus, the velocity vectors of curves through $p = \lambda(0)$ all lie in the same plane called the tangent plane at $z(p)$. Also, note that the velocity vector, $(z\circ\lambda)'$ depends only on the velocity vector, $\lambda'(0)$, and not on the curve λ. Thus, z induces a linear map (called the differential dz) that takes vectors at $p = \lambda(0)$ to vectors in the tangent plane at $z(p)$. This differential is a similarity that multiples all length by $\Delta(p)$ and thus preserves angles. Also, the distortion of z along λ is also $\Delta(p)$.

DEFINITION. In the above situation we call $\Delta(p)$ the *distortion of the map z at the point p* and denote it **dist$(z)(p)$.**

*PROBLEM 16.2   UPPER HALF PLANE IS MODEL OF ANNULAR HYPERBOLIC PLANE

We were able to prove in Problem **5.1** that there are reflections about the radial geodesics, but only assumed (based on our physical experience with physical models) the existence of other geodesics and reflections through them. To assist us in looking at transformations of the annular hyperbolic space (with radius ρ) we use the upper half plane model. If f is a transformation taking the upper half plane R^{2+} to itself, then from the diagram

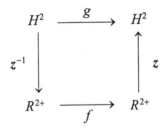

we see that $g = z \circ f \circ z^{-1}$ is a transformation from the annular hyperbolic plane to itself. We call g the transformation of H^2 that *corresponds* to f. We will call f an *isometry of the upper half plane model* if the corresponding g is an isometry of the annular hyperbolic plane. To show that g is an isometry, you must show that the transformation $g = z \circ f \circ z^{-1}$ preserves distances. Remember that distance along a curve is equal to the integral of the speed along the curve. Thus, it is enough to check that the distortion of g at each point is equal to 1. Before we do this we must first show

   a. *The distortion of an inversion i_C with respect to a circle Γ at a point P, which is a distance s from the center C of Γ, is equal to r^2/s^2, where r is the radius of the circle.* See Figure 16.2.

Hint: Because the inversion is conformal, the distortion is the same in all directions. Thus check the distortion along the ray from C, the center of circle, through P. The distance along this ray of an arbitrary point can be parametrized by $t \mapsto ts$. Use the definition of distortion given in Problem **16.1**.

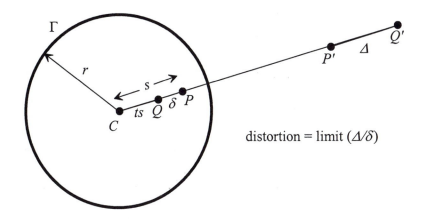

Figure 16.2   Distortion of an inversion

b. *Let f be the inversion in a circle whose center is on the x-axis. Show that f takes R^{2+} to itself and that $g = z \circ f \circ z^{-1}$ has distortion 1 at every point and is thus an isometry.*

OUTLINE OF A PROOF:

1. Note that each of the maps z, z^{-1}, f are conformal and have at each point a distortion that is the same for all curves at that point. If dist$(k)(p)$ denotes the distortion of the function k at the point p, then argue that dist$(g)(p) = $ dist$(z^{-1})(p) \times $ dist$(f) \times $ dist(z).

2. If $z(a,b) = p$, then show (using **5.1c**) that dist$(z^{-1})(p) = b/\rho$, where ρ is the radius of the annuli.

3. Show (using Part **a**) that dist$(f)(z^{-1}(p)) = r^2/s^2$, where r is the radius of the circle C which defines f and s is the distance from the center of C to (a,b).

4. Then show that dist$(z)(f(z^{-1}(p))) = \dfrac{\rho}{b\frac{r^2}{s^2}}$.

We call these inversions (or the corresponding transformations in the annular hyperbolic plane) *hyperbolic reflections*. We also call reflections through vertical half lines (corresponding to radial geodesics) hyperbolic reflections.

Now, you can prove that

   c. *If γ is a semicircle in the upper half plane with center on the x-axis or a straight half line in the upper half plane perpendicular to x-axis, then $z(\gamma)$ is a geodesic in the annular hyperbolic plane.*

Because of this, we say that such γ are *geodesics in the upper half plane model*. Because the compositions of two isometries is an isometry, we see immediately that any composition of inversions in semicircles (whose centers are on the x-axis) is an isometry in the upper half plane model (that is, the corresponding transformation in the annular hyperbolic plane is an isometry).

PROBLEM 16.3 PROPERTIES OF HYPERBOLIC GEODESICS

   a. *Any similarity of the upper half plane corresponds to an isometry of an annular hyperbolic. Such similarities must have their centers on the x-axis. (Why?)*

Look at the composition of inversions in two concentric semicircles.

   b. *If γ is a semicircle in the upper half plane with center on the x-axis, then there is an inversion (in another semicircle) that takes γ to a vertical line that is tangent to γ.*

Hint: An inversion takes any circle through the center of the inversion to a straight line (see Problem **14.2**).

Each of the following three parts is concerned with finding a geodesic. Each problem should be looked at in both the annular hyperbolic plane and in the upper half plane model. In a crocheted annular hyperbolic plane one can construct geodesics by folding much the same way you can on a piece of (planar) paper. Geodesics in the upper half plane model can be constructed using properties of circles and inversions (see Problem **14.2**). You will also find Part **b** very useful.

    c. *Given two points A and B in a hyperbolic plane there is a unique geodesic joining A to B; and there is an isometry that takes this geodesic to a radial geodesic (or vertical line in the upper half plane model).*

In the upper half plane model, construct a circle with center on the *x*-axis that passes through *A* and *B*. Then use Part **b**.

We use *AB* to denote the unique geodesic segment joining *A* to *B*.

    d. *Given a geodesic segment AB with endpoints points A and B in a hyperbolic plane there is a unique geodesic that is a perpendicular bisector of AB.*

Use appropriate folding in annular hyperbolic plane. In the upper half plane model, make use of the properties of a reflection through a perpendicular bisector.

    e. *Given an angle ∠ABC in a hyperbolic plane, there is a unique geodesic that bisects the angle.*

In the upper half plane model, again use the properties of a reflection through the bisector of an angle.

    f. *Any two geodesics on a hyperbolic plane either intersect, are asymptotic, or have a common perpendicular.*

Look at two geodesics in the upper half plane model that do not intersect in the upper half plane nor on the bounding *x*-axis.

PROBLEM **16.4** HYPERBOLIC IDEAL TRIANGLES

In Problem **7.2** we investigated the area of triangles in a hyperbolic plane. In the process we looked at ideal triangles and 2/3-ideal triangles. We can look more analytically at the ideal triangles. It is impossible to picture the whole of an ideal triangle in an annular hyperbolic plane, but it is easy to picture ideal triangles in the upper half plane model. In the upper half plane model an ***ideal triangle*** is a triangle with all three verti-ces either on the *x*-axis or at infinity. See Figure 16.3.

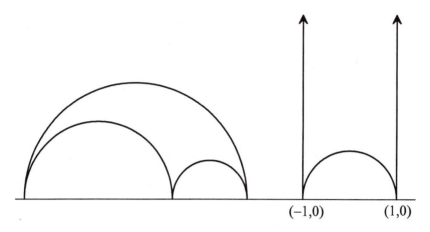

(−1,0)   (1,0)

Figure 16.3   Ideal triangles in the upper half plane model

At first glance it appears that there must be many different ideal triangles. However,

    a. *Prove that all ideal triangles on the same hyperbolic plane are congruent.*

Review your work on Problem **14.2**. Perform an inversion that takes one of the vertices to infinity and the two sides from that vertex to vertical lines. Then apply a similarity to the upper half plane taking the standard ideal triangle with vertices (−1,0), (0,1), and ∞.

    b. *Show that the area of an ideal triangle is $\pi\rho^2$.* (Remember this ρ is the radius of the annuli.)

Hint: Because the distortion dist$(z)(a,b)$ is ρ/b, the desired area is

$$\int_{-1}^{1} \int_{\sqrt{1-x^2}}^{\infty} \left(\frac{\rho}{y}\right)^2 dy\, dx.$$

We now picture in Figure 16.4 **2/3-*ideal triangles*** in the upper half plane model.

    c. *Prove that all 2/3-ideal triangles with angle θ are congruent and have area $(\pi-\theta)\rho^2$.*

Show, using Problem **14.2**, that all 2/3-ideal triangles with angle θ are congruent to the standard one at the right of Figure 16.4 and show that the area is the double integral

$$\int_{-1}^{\cos\theta} \int_{\sqrt{1-x^2}}^{\infty} \left(\frac{p}{y}\right)^2 dy\,dx.$$

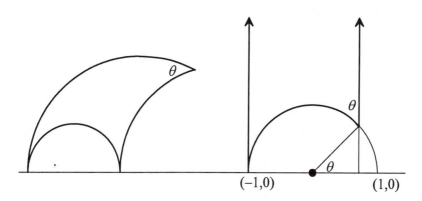

Figure 16.4   2/3-ideal triangles in the upper half plane model

PROBLEM **16.5** POINCARÉ DISK MODEL

You showed in Problem **16.1c** that the coordinate map x from a hyperbolic plane to the upper half plane preserves angles (is conformal); this we called the *upper half plane model*. Now we will study other models of the hyperbolic plane.

Let $z\colon H^2 \to R^{2+}$ be the coordinate map defined in Problem **16.2** that defines the upper half plane model. We will now transform the upper half plane model to a disk model that was first discussed by Poincaré in 1882.

a. *Show that any inversion through a circle whose center is in the lower half plane (that is, $y < 0$) will transform the upper half plane onto an open (without its boundary) disk. Show that the hyperbolic geodesics in the upper half plane are transformed by this inversion into circular arcs (or line segments) perpendicular to the boundary of the disk.*

Review the material on inversions discussed in Problem **14.2**.

b. *If w: $D^2 \to R^{2+}$ is the inverse of a map from the upper half plane to a (open) disk from Part **a**, then show that the composition*

$$z \circ w : D^2 \to H^2$$

is conformal. We call this the (***Poincaré***) ***disk model***, after Henri Poincaré (1854–1912, French).

Review the material on inversions in **14.2** and on the upper half plane model in **16.2**.

c. *Show that any inversion through a circular arc (or line segments) perpendicular to the boundary of D^2 takes D^2 to itself. Show that these inversions correspond to isometries in the (annular) hyperbolic plane.* Thus, we call these circular arcs (or line segments) ***hyperbolic geodesics*** and call the inversions ***hyperbolic reflections*** in the Poincaré disk model.

Review Problem **16.2**.

See Figure 16.5 for a drawing of geodesics and a triangle in the Poincaré disk.

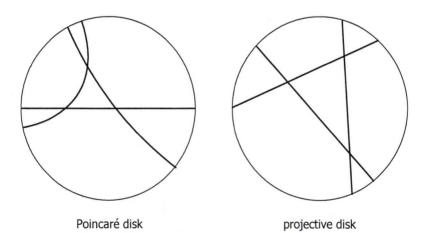

Poincaré disk   projective disk

Figure 16.5   Geodesics and a triangle

PROBLEM **16.6**   PROJECTIVE DISK MODEL

Let D^2 be the disk model of a hyperbolic plane and assume its radius is 2. Then place a sphere of radius 1 tangent to the disk at its center. Call this point of tangency the South Pole S. See Figure 16.6.

Now let s be the stereographic projection from the sphere to the plane containing D^2, and note that s(equator) is the boundary of D^2, and thus s takes the Southern Hemisphere onto the D^2. Now let h be the projection of the Southern Hemisphere onto the disk, B^2, of radius 1.

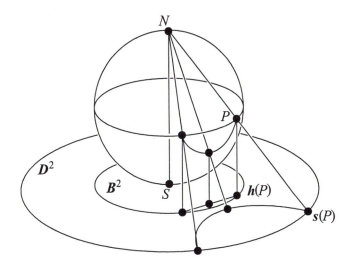

Figure 16.6   Obtaining the projective disk model from the Poincaré disk model

Show that the mapping $h \circ s^{-1}$ takes D^2 to B^2 and takes each circle (or diameter) of D^2 to a (straight) cord of B^2. Thus

$$h \circ s^{-1} \circ (z \circ w)^{-1}$$

is a map from the hyperbolic plane to B^2, which takes geodesics to straight line segments (cords) in B^2.

We call this the *projective disk model*, but it is also in the literature called the *Beltrami/Klein model* or the *Klein model*, named after

Eugenio Beltrami (1835–1900, Italian) who described the model in 1868, and Felix Klein (1849–1925, German) who fully developed it in 1871.

See Figure 16.5 for a drawing of geodesics and a triangle in the projective disk model.

Chapter 17

Geometric 2-Manifolds and Coverings

The concept "two-dimensional manifold" or "surface" will not
be associated with points in three-dimensional space; rather it
will be a much more general abstract idea.
— Hermann Weyl (1913)

There are clearly a large variety of different surfaces around in our
experiential world. The study of the geometry of general surfaces is the
subject of Differential Geometry. In this chapter we will study surfaces
that (like cylinders and cones with the cone point removed) are locally
the same (isometric) to either the plane, a sphere, or a hyperbolic plane.
We study these because their geometry is simpler and closely related to
the geometry we have been studying of the plane, spheres, and hyperbolic planes. In addition, the study of these surfaces will lead us to the
study of the possible global shapes of our physical universe.

There are no prerequisites for this chapter from after Chapter 7, but
some of the ideas may be difficult the first time around. Problems **17.2**,
17.4, and **17.5**, and **17.7** are what is needed from this chapter before you
study Chapter 22 (*3-Manifolds — Shape of Space*); the other Problems
can be skipped.

A *geometric 2-manifold* is a connected space that locally is isometric to either the (Euclidean) plane, a sphere, or a hyperbolic plane. The
surface of a cylinder (no top or bottom and indefinitely long) and a cone
(with the cone point removed) are examples of geometric 2-manifolds.
We use the term "manifold" here instead of "surface" because we usually think of surfaces as sitting extrinsically in 3-space. Here we want to
study only the intrinsic geometry and thus any particular extrinsic

embedding does not matter. Moreover, we will study some geometric 2-manifolds (for example, the flat torus) that cannot be (isometrically) embedded in 3-space. We ask what is the intrinsic geometric experience on geometric 2-manifolds of a 2-dimensional bug. How will the bug view geodesics (intrinsically straight lines) and triangles? How can a bug on a geometric 2-manifold discover the global shape of its universe? These questions will help us as we think about how we as human beings can think about our physical universe, where *we* are the bugs.

This chapter will only be an introduction to these ideas. For a geometric introduction to Differential Geometry see [**DG:** Henderson]. For more details about geometric 2-manifolds see [**DG:** Weeks] and Chapter 1 of [**DG:** Thurston]. For the classification of (triangulated) 2-manifolds see the recent [**Tp:** Francis & Weeks], which contains an accessible proof due to John H. Conway.

*PROBLEM 17.1 GEODESICS ON CYLINDERS AND CONES

In Problem **4.1**, we have already studied two examples of geometric 2-manifolds — cylinders and cones (without the cone point). Because these surfaces are locally isometric to the Euclidean plane, these type of geometric manifolds are called *flat* (or *Euclidean*) *2-manifolds*. It would be good at this point for you to review what you know from Chapter 4 about cylinder and cones.

Now we will look more closely at long geodesics that wrap around on a cylinder or cone. Several questions have arisen.

> **a.** *How do we determine the different geodesics connecting two points? How many are there? How does it depend on the cone angle? Is there always at least one geodesic joining each pair of points? How can we justify our conjectures?*

> **b.** *How many times can a geodesic on a cylinder or cone intersect itself? How are the self-intersections related to the cone angle? At what angle does the geodesic intersect itself? How can we justify these relationships?*

SUGGESTIONS

Here we offer the tool of covering spaces, which may help you explore these questions. The method of "coverings" is so named because it utilizes layers (or sheets) that each "cover" the surface. We will first start with a cylinder because it is easier, and then move on to a cone.

n-Sheeted Coverings of a Cylinder

To understand how the method of coverings works, imagine taking a paper cylinder and cutting it axially (along a vertical generator) so that it unrolls into a plane. This is probably the way you constructed cylinders to study this problem before. The unrolled sheet (a portion of the plane) is said to be a *1-sheeted covering* of the cylinder. See Figure 17.1. If you marked two points on the cylinder, *A* and *B*, as indicated in the figure, when the cylinder is cut and unrolled into the covering, these two points become two points on the covering (which are labeled by the same letters in the figure). The two points on the covering are said to be *lifts* of the points on the cylinder.

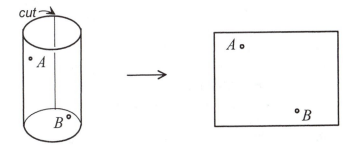

Figure 17.1   A 1-sheeted covering of a cylinder

Now imagine attaching several of these "sheets" together, end-to-end. When rolled up, each sheet will go around the cylinder exactly once — they will each cover the cylinder. (Rolls of toilet paper or paper towels give a rough idea of coverings of a cylinder.) Also, each sheet of the covering will have the points *A* and *B* in identical locations. You can see this (assuming the paper thickness is negligible) by rolling up the coverings and making points by sticking a sharp object through the cylinder. This means that all the *A*'s are coverings of the same point on the cylinder and all the *B*'s are coverings of the same point on the cylinder. We just have on the covering several representations, or *lifts*, of each point on the cylinder. Figure 17.2 depicts a 3-sheeted covering space for a cylinder and six geodesics joining *A* to *B*. (One of them is the most direct path from *A* to *B* and the others spiral once, twice, or three times around the cylinder in one of two directions.)

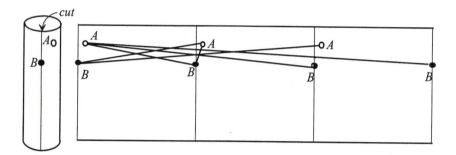

Figure 17.2   A 3-sheeted covering space for a cylinder

We could also have added more sheets to the covering on either the right or left side. You can now roll these sheets back into a cylinder and see what the geodesics look like. Remember to roll it up so that each sheet of the covering completely covers the cylinder — all of the vertical lines between the coverings should lie on the same generator of the cylinder. Note that if you do this with ordinary paper, part or all of some geodesics will be hidden, even though they are all there. It may be easier to see what's happening if you use transparencies.

This method works because straightness is a local intrinsic property. Thus, lines that are straight when the coverings are laid out in a plane will still be straight when rolled into a cylinder. Remember that bending the paper does not change the intrinsic nature of the surface. Bending only changes the curvature that we see extrinsically. It is important to always look at the geodesics from the bug's point of view. The cylinder and its covering are locally isometric.

Use coverings to investigate Problem **17.1** on the cylinder. The global behavior of straight lines may be easier to see on the covering.

n-Sheeted (Branched) Coverings of a Cone

Figure 17.3 shows a 1-sheeted covering of a cone. The sheet of paper and the cone are locally isometric except at the cone point. The cone point is called a **branch point** of the covering. We talk about lifts of points on the cone in the same way as on the cylinder. In Figure 17.3 we depict a 1-sheeted covering of a 270° cone and label two points and their lifts.

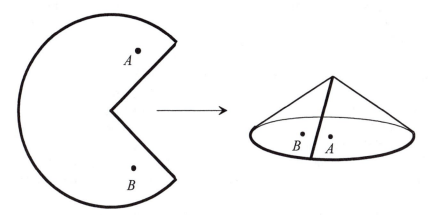

Figure 17.3   1-sheeted covering of a 270° cone

A 4-sheeted covering space for a cone is depicted in Figure 17.4. Each of the rays drawn from the center of the covering is a lift of a single ray on the cone. Similarly, the points marked on the covering are the lifts of the points *A* and *B* on the cone. In the covering there are four segments joining a lift of *A* to different lifts of *B*. Each of these segments is the lift of a different geodesic segment joining *A* to *B*.

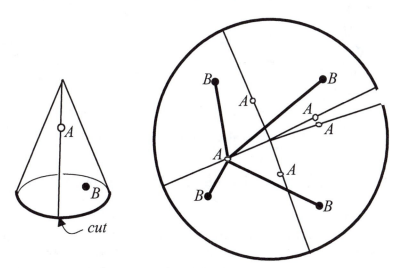

Figure 17.4   4-sheeted covering space for a 89° cone

Think about ways that the bug can use coverings as a tool to expand its exploration of surface geodesics. Also, think about ways you can use coverings to justify your observations in an intrinsic way. It is important to be precise; you don't want the bug to get lost! Count the number of ways in which you can connect two points with a straight line and relate those countings with the cone angle. Does the number of straight paths only depend on the cone angle? Look at the 450° cone and see if it is always possible to connect any two points with a straight line. **Make paper models!** It is not possible to get an equation that relates the cone angle to the number of geodesics joining every pair of points. However, it is possible to find a formula that works for most pairs.

Make covering spaces for cones of different size angles and refine the guesses you have already made about the numbers of self-intersections.

In studying the self-intersections of a geodesic l on a cone, it may be helpful for you to consider the ray R such that the line l is perpendicular to it. (See Figure 17.5.) Now study one lift of the geodesic l and its relationship to the lifts of the ray R. Note that the seams between individual wedges are lifts of R.

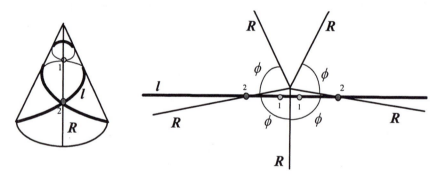

Figure 17.5   Self-intersections on a cone with angle ϕ

PROBLEM 17.2   FLAT TORUS AND FLAT KLEIN BOTTLE

FLAT TORUS

Another example of a flat (Euclidean) 2-manifold is provided by a video game that was popular a while ago. A blip on the video screen representing a ball travels in a straight line until it hits an edge of the screen. Then, the blip reappears traveling parallel to its original direction from a point at the same position on the opposite edge. Is this a representation of some surface? If so, what surface? First, imagine rolling the screen into a tube where the top and bottom edges are glued. (Figure 17.6.) This is a representation of the screen as a 1-sheeted covering of the cylinder. A blip on the screen that goes off the top edge and reappears on the bottom is the lift of a point on the cylinder that travels around the cylinder crossing the line that corresponds to the joining of the top and bottom of the screen.

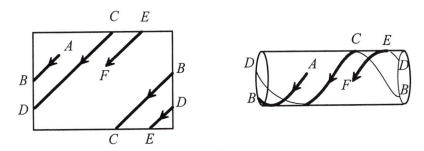

Figure 17.6   1-sheeted covering of cylinder and flat torus

Now, let us further imagine that the cylinder can be stretched and bent so that we can glue the two ends to make a torus. Now the screen represents a 1-sheeted covering of the torus. If the blip goes off on one side and comes back on the other at the same height, this represents the lift of a point moving around the torus and crossing the circle that corresponds to the place where the two ends of the cylinder are joined. The possible motions of a point on the torus are represented by the motions on the video screen!

You can't make a model in 3-space of a flat torus from a flat piece of paper without distorting it. Such a torus is called a ***flat torus***. It is best not to call this a "surface", because there is no way to realize it isometrically in 3-space and it is not the surface of anything. But the question of whether or not you can make an isometric model in 3-space is not important — the point is that the gluings in Figure 17.6 intrinsically define a flat 2-manifold.

If you distort the cylinder in Figure 17.6 in 3-space you can get the torus pictured in Figure 17.7. This is not a geometric 2-manifold because the original flat (Euclidean) geometry has been distorted and it is also not exactly either spherical or hyperbolic.

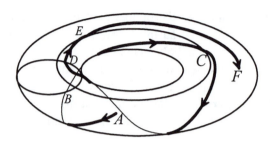

Figure 17.7   Non-flat torus

a. *Show that the flat torus is locally isometric to the plane and thus, is a geometric 2-manifold, in particular, a flat (Euclidean) 2-manifold.*

Note that each point on the interior of an edge of the screen is the lift of a point that has another lift on the opposite edge. Thus, a lift of a neighborhood of that is in two pieces (one near each of the two opposite edges). What happens at the four corners of the computer screen (which are lifts of the same point)?

The torus in Figure 17.7 and the flat torus are related in that there is a continuous one-to-one mapping from either to the other. We say that they are *homeomorphic*, or *topologically equivalent*. We can further express this situation by saying that the torus in Figure 17.7 and the flat torus are both *topological tori*.

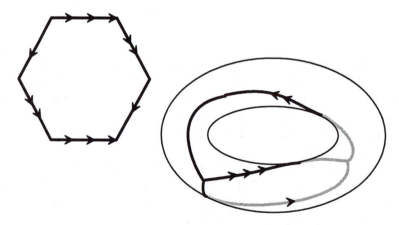

Figure 17.8   Flat torus from a hexagon

There is another representation of the flat torus based on a hexagon. Start with a regular hexagon in the plane and glue opposite sides as indicated in Figure 17.8.

> ***b.** *Show that gluing the edges of the hexagon as in Figure 17.8 forms a flat 2-manifold (called the **hexagonal torus**).*

*Flat Klein Bottle

Now we describe a related geometric 2-manifold, traditionally called a *flat Klein bottle,* named after Felix Klein (1849–1925, German). Imagine the same video screen, again with a traveling blip representing a ball that travels in a straight line until it hits an edge of the screen. When it hits the top edge then the blip proceeds exactly the same as for the flat torus (traveling parallel to its original direction from a point at the same position on the opposite edge). However, when the blip hits a vertical edge of the screen it reappears on the opposite edge but in the *diametrically opposite* position and travels in a direction with slope that is the negative of the original slope. (See Figure 17.9.) As before, imagine rolling the screen into a tube where the top and bottom edges are joined. This is again a representation of the screen as a 1-sheeted covering of the cylinder.

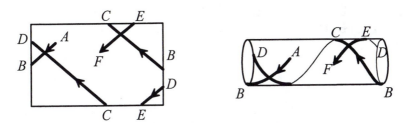

Figure 17.9   1-sheeted covering of a flat Klein bottle

A blip on the screen that goes off the top edge and reappears on the bottom is the lift of a point on the cylinder that travels around the cylinder crossing the line that corresponds to the joining of the top and bottom of the screen.

Now, let us further imagine that the cylinder can be stretched and bent so that we can join the two ends to make a topological Klein bottle, which is not a geometric 2-manifold. See Figure 17.10. Now the screen represents a 1-sheeted covering of the Klein bottle. The possible motions of a point on the Klein bottle are represented by the motions on the video screen!

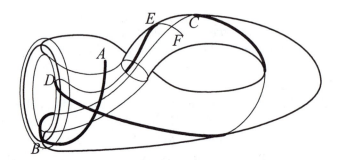

Figure 17.10   Topological Klein bottle

You also can't make an isometric model in 3-space of a flat Klein bottle without distorting it and having self-intersections. But the gluings in Figure 17.9 define intrinsically a flat 2-manifold.

> ***c.** *Show that the flat Klein bottle is locally isometric to the plane and thus is a geometric 2-manifold, in particular, a flat (Euclidean) 2-manifold.*

Note that the four corners of the video screen are lifts of the same point and that a neighborhood of this point has 360° — that is, 90° from each of the four corners.

It can be shown that

THEOREM 17.2. *Flat tori and flat Klein bottles are the only flat (Euclidean) 2-manifolds that are finite and geodesically complete (every geodesic can be extended indefinitely). See* Relations to Differential Geometry, *last section in Chapter 4.*

For a detailed discussion, see [**DG**: Thurston], pages 25–28. For a more elementary discussion see [**DG**: Weeks], Chapters 4 and 11.

Note that a finite cylinder is not geodesically complete; and if it is extended indefinitely then it is geodesically complete but not finite. A cone with the cone point is not a flat manifold at the cone point; with the cone point removed the cone is not geodesically complete.

Note that we get a flat tori for each size rectangle in the plane. These flat tori are different geometrically because there are different distances around the tori. However, topologically they are all the same as (homeomorphic to) the surface of a doughnut.

Note that if you move a right hand glove (which we stylize by 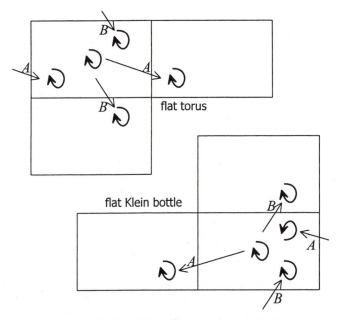) around the flat torus it will always stay right-handed; however, if you move it around the flat Klein bottle horizontally it will become left-handed. See Figure 17.11. We describe these phenomena by saying that the flat torus is *orientable* and a Klein bottle is *non-orientable*.

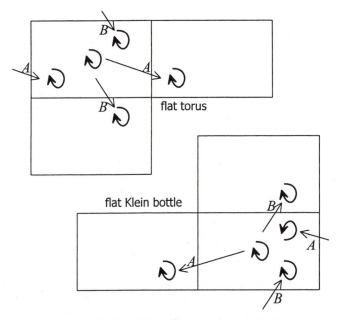

Figure 17.11   Orientable and non-orientable

*PROBLEM **17.3** UNIVERSAL COVERING OF FLAT 2-MANIFOLDS

a. *On a flat torus or flat Klein bottle, how do we determine the different geodesics connecting two points? How many are there? How can we justify our conjectures?*

Look at straight lines in the universal coverings introduced below.

b. *Show that some geodesics on the flat torus or flat Klein bottle are closed curves (in the sense that they come back and continue along themselves like great circles), though possibly self-intersecting. How can you find them?*

Look in the universal coverings introduced below.

c. *Show that there are geodesics on the flat torus and flat Klein bottle that never come back and continue along themselves.*

Look at the slopes of the geodesics found in Part **b**.

The geodesics found in Part **c** can be shown to come arbitrarily close to *every* point on the manifold. Such curves are said to be ***dense*** in the manifold.

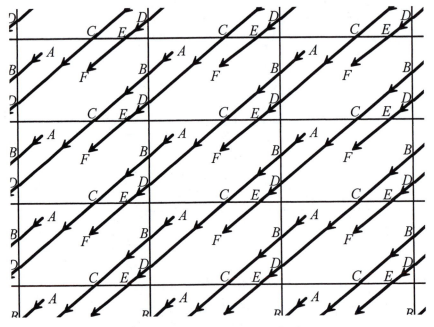

Figure 17.12   Universal covering of a flat torus

SUGGESTIONS

I suggest that you use coverings just as you did for cones and cylinders. The difference is that in this case the sheets of the coverings extend in two directions. See Figure 17.12 for a covering of the flat torus. If this covering is continued indefinitely in all directions then the whole plane covers the flat torus with each point in the torus having infinitely many lifts. When a covering is the whole of either the (Euclidean) plane, a sphere, or a hyperbolic plane it is called *the universal covering*. See Figure 17.13 for a universal covering of a flat Klein bottle. These coverings are called "universal" because there are no coverings of the plane, spheres, or hyperbolic planes that have more than one sheet — see the next section for a discussion of this for a sphere.

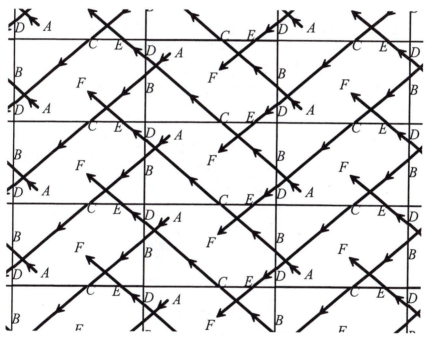

Figure 17.13   Universal covering of the flat Klein bottle

PROBLEM **17.4**   SPHERICAL **2-MANIFOLDS**

Start by considering another version of the video screen as depicted in Figure 17.14. Imagine the same video screen, again with a traveling blip representing a ball that travels in a straight line until it hits an edge of the screen. Now when the blip reaches *any* edge of the screen it reappears on the opposite edge but in the *diametrically opposite* position, and travels in a direction with slope that is the negative of the original slope. (See Figure 17.14.)

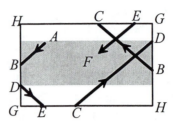

Figure 17.14   This is not a geometric 2-manifold

a. *Show that the situation in Figure 17.14 does not represent a geometric 2-manifold because the corners represent two cone points with cone angles* 180°.

Cut around the corners marked *G* and tape together the edges as indicated.

If you restrict yourself to the shaded strip in Figure 17.14 and identify the left and right edges as indicated, then you obtain a ***Möbius strip***, named after August Möbius (German, 1790–1868). You may have seen this surface before — if not, you should be sure to construct one from a (preferably long) strip of paper. The Möbius strip fails to be a geometric manifold only because it has an edge (note that there is only one edge!); but it is an example of what is called a "geometric manifold with boundary."

The gluings (on all four edges) indicated in Figure 17.14 fail to produce a geometric 2-manifold because the interior angles are only 90°. It seems that we might get a geometric manifold if the interior angles were 180°. Thus, we need a quadrilateral with equal opposite sides and interior angles with 180°. There is no such quadrilateral in the plane; however, on the sphere there IS such a quadrilateral! See Figure 17.15.

What we have in Figure 17.15 is a hemisphere with each point of the bounding equator being glued to its antipode. In this way one gets what is called the (real) projective plane, often denoted **RP**^{2}.

b. *Show that a projective plane is a spherical 2-manifold that contains a Möbius strip.*

Examine the neighborhood of a point of the projective plane that comes from the bounding equator.

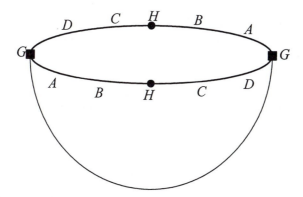

Figure 17.15   Gluings on a hemisphere producing a projective plane

c. *What are the geodesics on a projective plane?*

It is clear that the geodesics come from half great circles in the hemi-sphere, but what happens as one of these half great circle is crossing the equator that is glued?

If you cut out and remove a lune from a sphere and then join together the two edges of the lune you have what can reasonably be called a spherical cone. See Figure 17.16. Note that at least some of these spherical cones have a shape similar to an American football.

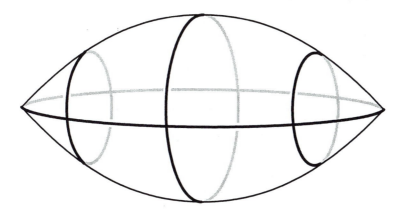

Figure 17.16   A spherical cone

By pasting together several (equal radius) spheres with the same lune removed you can get multiple-sheeted branched coverings of a spherical cone. A spherical cone with the two cone points removed is a finite spherical 2-manifold but (as with ordinary cones) it is not geodesically complete.

> **d.** *Show that the spherical cones as described above are spherical 2-manifolds, if you remove the two cone points.*

> **e.** *Identify the geodesics on a spherical cone with cone angle 180°* (that is, you remove from the sphere a lune with angle 360° – 180° = 180°). *What happens with other cone angles?*

Look at the great circles in the sphere minus the lune before its edges are joined to produce the spherical cone.

It can be shown that

THEOREM 17.4. *Spheres and projective planes are the only spherical 2-manifolds which are finite and geodesically complete (every geodesic can be extended indefinitely).*

For a detailed discussion, see [**DG**: Thurston], pages 25–18. For a more elementary discussion see [**DG**: Weeks], Chapters 4 and 11.

*COVERINGS OF A SPHERE

There is no way to construct a covering of a sphere that has more than one sheet unless the covering has some "branch points." A **branch point** on a covering is a point such that every neighborhood (no matter how small) surrounding the point contains at least two lifts of some point. In any covering of a cone with more than one sheet, the lift of the cone point is a branch point as you can see in Figure 17.17.

Notice that the coverings of a cylinder and a flat torus have no branch points. For a sphere the matter is very different — any covering of a sphere will have a branch point. You can see this if you try to construct a cover by slitting two spheres as depicted in Figure 17.18 and then sticking the two together along the slit. The ends of the slit would become branch points. This topic may be explored further in textbooks on geometric or algebraic topology.

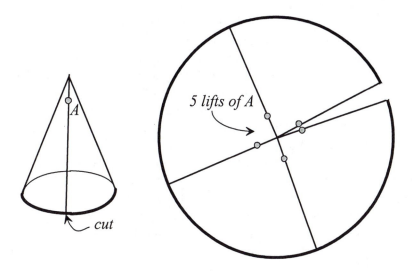

Figure 17.17   Covering space of a cone has branch points

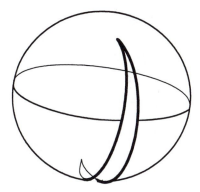

Figure 17.18   Covering space of a sphere has branch points

In fact, any surface that has no (non-branched) coverings and that is bounded and without an edge can be continuously deformed (without tearing) into a round sphere. The surfaces of closed boxes and of footballs are two examples. A torus is bounded and without an edge, but it cannot be deformed into a sphere. A cylinder also cannot be deformed into a sphere, and a cylinder either has an edge or (if we imagine it as extending indefinitely) it is unbounded.

A 3-dimensional analog of this situation arises from a famous, long-unsolved problem called the *Poincaré Conjecture.* The analog of a surface is called a 3-*dimensional manifold,* a space locally like Euclidean 3-space, (in the same sense that a surface is locally like the plane). The 3-sphere, which we will study in Chapter 20, is a 3-dimensional manifold. Also, in Chapter 22 we note that our physical 3-dimensional universe is a 3-dimensional manifold. Poincaré (1854–1912, French) conjectured that any 3-dimensional manifold that has no (non-branched) coverings and that is bounded and without boundary must be homeomorphic to a 3-dimensional sphere S^3. For the past 80 years, numerous mathematicians have tried to decide whether Poincaré's conjecture is true. So far, no one has succeeded. On May 24, 2000, the Clay Mathematics Institute announced (www.claymath.org) that it was offering a $1,000,000 prize for a solution of the Poincaré Conjecture! See Chapter 22 and [**SE**: Hilbert] and [**DG**: Weeks] for more discussion of 3-dimensional manifolds and the 3-dimensional sphere.

PROBLEM **17.5** HYPERBOLIC MANIFOLDS

Now, is it possible to make a two-holed torus (sometimes called an anchor ring, or the surface of a two-holed donut) into a geometric 2-manifold? See Figure 17.19.

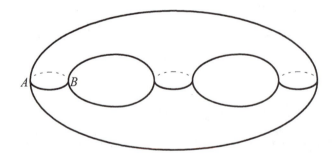

Figure 17.19   Two-holed torus

Note that as it is pictured in Figure 17.19 the two-holed torus is definitely **not** a geometric 2-manifold because the intrinsic geometry is not the same at every point — for example, points *A* and *B*. But can we distort the geometry so that the surface is a geometric 2-manifold?

Imagine cutting the two-holed torus along the four loops emanating from the point *P*, as indicated in Figure 17.20. You will get a distorted octagon with 45° (= 360°/8) interior angles at each vertex. This distorted octagon is topologically equivalent to a regular planar octagon. Walking around the point P we find the gluings as indicated in Figure 17.20. Be sure you understand how the gluings on the octagon where determined from the loops on the two-holed torus. If you glue the edges of the regular octagon together as indicated in Figure 17.20, you will get a version of the two-holed torus that is geometrically the plane except at the (one) vertex. (*Why? What will a neighborhood of the vertex look like?*) In the plane all octagons have the same interior angle sum. But, in the hyperbolic plane, regular octagons have different angles. In fact, we can find such an octagon in the hyperbolic plane with 45° interior angles. (See Figure 17.21). If we glue the edges of this octagon as indicated then we will get a hyperbolic 2-manifold.

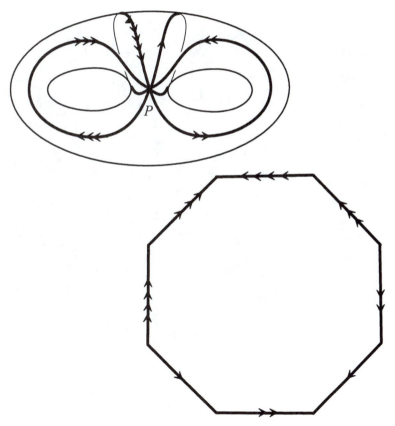

Figure 17.20   Cutting a two-holed torus

To see that there is such an octagon, imagine placing a small (regular) octagon on the hyperbolic plane. Because the octagon is small, its interior angles must be very close to the interior angles of an octagon in the (Euclidean) plane. Because the exterior angle of an planar octagon must be 360°/8, the interior angle must be 180° – (360°/8) = 135°. Now let the small octagon grow, keeping it always regular. From Problem **7.2** (remembering that an octagon can be divided into triangles as in Figure 17.21), we conclude that the interior angles of the octagon will decrease in size until, if we let the vertices go to infinity, the angles would decrease to zero. Somewhere between 135° and 0° the interior angles will be the desired 45°.

    a. *What is the area of the hyperbolic octagon with 45° interior angles?*

Figure 17.21   Hyperbolic octagon with 45° angles

    b. *Why is the two-holed torus obtained from a hyperbolic octagon with 45° interior angles a hyperbolic 2-manifold? What is its area?*

It is not easy to determine the geodesics on a hyperbolic 2-manifold. Some discussion about these issues is contained in William Thurston's *Three-Dimensional Geometry and Topology*, Volume 1 [**DG**: Thurston] and will presumably be continued in the forthcoming Volume 2.

There are other ways of making a two-holed torus into a geometric 2-manifold but it is always a *hyperbolic* 2-manifold. However, there are many different hyperbolic structures for a two-holed torus; for example, look in Figure 17.22 for a different way to represent the two-holed torus — this time as the gluing of the boundary of a dodecagon with 90° interior angles.

   c. *Follow the steps above to check that Figure* 17.22 *leads to the representation of a two-holed torus as a dodecagon* (with 90° interior angles) *with gluing on the boundary and thus to a hyperbolic 2-manifold. What is its area?*

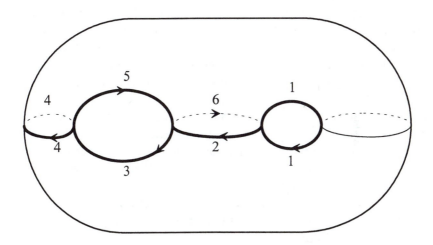

Figure 17.22   Cutting two-holed torus into a dodecagon

You should have found the same area in Parts **b** and **c**. In fact, we (you!) will show in the next problem that *any hyperbolic geometric manifold structure on a two-holed torus has the same area.*

All of the above discussion can be extended in straightforward ways to tori with 3, 4, and more holes.

Problem 17.6   Area, Euler Number, and Gauss–Bonnet

Now is a good time to go back and review the material on area and holonomy in Problems **7.1–7.4a**. Recall that you showed that the area of any polygon is

$$K \, Area(\Gamma) = [\, \Sigma \beta_i - (n-2)\pi \,],$$

where K is the Gaussian curvature equal to $1/\rho^2$ for a sphere and $-1/\rho^2$ for a hyperbolic plane of radius ρ, and $\Sigma \beta_i$ is the sum of the interior angles, and n is the number of edges.

We will now use those results to study the area of geometric manifolds. All of the geometric manifolds that we have described above have cell-divisions. A **0-*cell*** is a point we usually call a *vertex*. A **1-*cell*** is a straight line segment we usually call an *edge* (the edge need not be straight but in most of our applications it will be). A **2-*cell*** is a polygon that is usually called a *face*. We say that a geometric manifold has a ***cell-division*** if it is divided into cells so that every edge has its boundary consisting of vertices and every face has its boundary divided into edges and vertices and two cells only intersect on their boundaries. We call the cell-division a ***geodesic cell-division*** if all the edges are geodesic segments, and thus the faces are polygons. For example, in Figure 17.22 we have a two-holed torus divided into one face, six edges, and three vertices. Note that some edges have only one vertex and thus form a loop (circle). If we make this two-holed torus into a hyperbolic manifold (by using a regular hyperbolic dodecagon with 90° angles) then the cell-division is a geodesic cell-division.

Suppose we have a geodesic cell-division of a geometric manifold M into f faces (2-cells) where the j-th face has n_j edges. We can calculate the area of M as follows:

$K \, area(M) =$

$= K\{area(\text{1-st face})\} + K\{area(\text{2-nd face})\} + \dots + K\{area(f\text{-th face})\}$

$= \{\Sigma\beta_i \,(\text{1-st face}) - (n_1 - 2)\pi\} + \dots + \{\Sigma\beta_i \,(f\text{-th face}) - (n_f - 2)\pi\}$

$= \{\Sigma\beta_i \,(\text{1-st face}) + \dots + \Sigma\beta_i \,(f\text{-th face})\} - \{(n_1 - 2)\pi + \dots + (n_f - 2)\pi\}$

$= \{\text{sum of all the angles}\} - (n_1 + n_2 + \dots + n_f)\pi + (2 + 2 + \dots + 2)\pi$

Fill in the steps to prove:

a. *If a geometric manifold M has a geodesic cell-division then*

$$K \text{ area } (M) = 2\pi \, (v - e + f),$$

where the cell-division has v vertices and e edges and f faces. Check that this agrees with your results in Problems **17.2** *and* **17.4**.

The quantity $(v - e + f)$ is called the ***Euler number*** (or sometimes the ***Euler characteristic***). This quantity is named after the mathematician Leonhard Euler (1707–1783) who was born and educated in Basel, Switzerland, but worked in St. Petersburg and Berlin. It follows directly from Part **a** and what we know about area and curvature that

b. *The Euler number of any geodesic cell-division of a sphere must be 2. The Euler number of any geodesic cell-division of a projective plane must be 1. The Euler number of any geodesic cell-division of a flat (Euclidean) 2-manifold must be equal to 0. The Euler number of any geodesic cell-division of a hyperbolic 2-manifold must be negative.*

We see from Part **b** that in the cases of a sphere, a projective plane, or a flat 2-manifold, the Euler number does not depend on the specific cell-division (with geodesic edges). So, we can talk about the ***Euler number of the sphere*** (= 2) and the ***Euler number of the torus*** or ***Klein bottle*** (= 0). We also saw above that the two different cell-divisions that were given of the two-holed torus have the same area and thus the same Euler number. Can we prove that for every hyperbolic 2-manifold the Euler number (and therefore the area) depends only on the topology and not on the particular cell-division? In fact,

Theorem 17.6a. *The Euler number of any cell-division of a 2-manifold depends only on the topology of the manifold and not on the specific cell-division. Furthermore, two 2-manifolds are homeomorphic if and only if they have the same Euler number and are either both orientable or both non-orientable.*

Proofs of this result are somewhat fussy and involve much of the foundational results of topology that were only developed in the twentieth century. See Imre Lakatos's *Proofs and Refutations* [**Ph:** Lakatos] for an accessible and interesting account of the long and complicated history and philosophy of the Euler number. Lakatos describes an imaginary class discussion about the Euler number in which the tortuous route students take toward a proof mirrors the actual route that mathematicians took. Other proofs are with different additional assumptions. For example, [**DG:** Thurston] (Propositions 1.3.10 and 1.3.12) gives an accessible proof assuming that the reader has some familiarity with vector fields on differentiable manifolds. In Sections 2.4 and 2.5 of [**Tp:** Blackett] there is a combinatorial-based proof that assumes the topological 2-manifold has some cell-division.

*TRIANGLES ON GEOMETRIC MANIFOLDS

Clearly, if on a flat (Euclidean) 2-manifold a triangle is contained in a region that is isometric to the plane, then the triangle is a planar triangle and has all the properties of a triangle in the plane. The same can be said about triangles in spherical and hyperbolic 2-manifolds. In fact it can be shown (see any topology text that deals with covering spaces) that

> **THEOREM 17.6B.** *If Δ is a triangle in a Euclidean [spherical, hyperbolic] 2-manifold, M, such that Δ can be shrunk to a point in the interior of Δ, then Δ (and its interior) can be lifted to the plane [sphere, hyperbolic plane] that is the universal covering space of M; and thus, Δ has all the same properties of a triangle in the plane [sphere, hyperbolic plane].*

It is natural to be uncomfortable using covering spaces, but covering spaces are a helpful tool for thinking intrinsically. Some triangles, even though they look strange extrinsically, will look like reasonable triangles for the bug. In Figure 17.23 we give an example of an extrinsically strange triangle that intersects itself, but which can be considered a normal triangle from an intrinsic point of view. In fact, it is a planar triangle. Such triangles have all the properties of plane triangles including SAS and ASA.

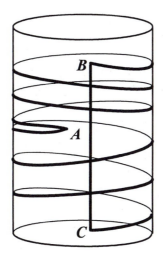

Unroll to a 3-sheeted cover, and ...

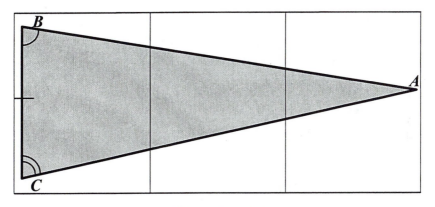

This is a triangle!

Figure 17.23   Think intrinsically

PROBLEM 17.7   CAN THE BUG TELL WHICH MANIFOLD?

Our physical universe is apparently a geometric 3-manifold. In Chapter 22 we will explore ways in which we (human beings) may be able to determine the global shape and size of our physical universe. But first we look at the situation of a 2-dimensional bug on a geometric 2-manifold in order to get some help for our 3-dimensional question.

a. *Suppose a 2-dimensional bug lives on a geometric 2-manifold M and suppose that M is the bug's whole universe. How can the bug determine intrinsically what the local geometry of its universe is? For this part you may imagine that the bug can crawl over the whole manifold and leave markers and make measurements.*

b. *How can the bug in Part* **a** *tell what the global shape of its universe is? For example, how can the bug tell (intrinsically) the difference between being on a flat torus and being on a hexagonal torus.*

c. *Suppose that the bug in Part* **a** *can only travel in a very small region of the manifold (so small that all triangles in the region are indistinguishable from planar triangles), but can see for very long distances. Can the bug still determine which geometric 2-manifold is its universe?*

You must assume that light travels on geodesics in the bug's universe (which is the case of light in our physical 3-dimensional universe). Or forget about light and think about vibrations that travel along the surface. Our 2-dimensional bugs have antennae with which they can receive these vibrations and through them "see" the surface. Image that there are bright stars on the 2-manifold that the bug can see; but remember that these stars must on the 2-manifold — we are taking an intrinsic view of 2-manifolds.

This is different from the situation on the earth where we have the extrinsic observation of stars (and sun and moon) to help us. The sun was used to measure the radius of the earth by Eratosthenes of Cyrene (Egyptian, 276–194 B.C.) and others. Mariners have used the sun, stars, and eclipses (and clocks) since early times for navigating on the oceans — for an interesting historical account of the problem of determining one's longitude on the earth see *Longitude* by Dava Sobel [**Z:** Sobel].

Chapter 18

GEOMETRIC SOLUTIONS OF QUADRATIC AND CUBIC EQUATIONS

Whoever thinks algebra is a trick in obtaining unknowns has thought it in vain. No attention should be paid to the fact that algebra and geometry are different in appearance. Algebras (jabbre and maqabeleh) are geometric facts which are proved by Propositions Five and Six of Book Two of [Euclid's] Elements.

— Omar Khayyam, a paper [**AT:** Khayyam (1963)]

In this chapter we will see how the results from Chapter 13 were used historically to solve equations. Quadratic equations were solved by "completing the square" — a real square. These in turn lead to conic sections and cube roots and culminate in the beautiful general method from Omar Khayyam which can be used to find all the real roots of cubic equations. Along the way we shall clearly see some of the ancestral forms of our modern Cartesian coordinates and analytic geometry. I will point out several inaccuracies and misconceptions that have crept into the modern historical accounts of these matters. But I urge you not to look at this only for its historical interest but also for the meaning it has in our present-day understanding of mathematics. This path is not through a dead museum or petrified forest; it passes through ideas that are very much alive and have something to say to our modern techno-logical, increasingly numerical, world.

PROBLEM 18.1   QUADRATIC EQUATIONS

Finding square roots is the simplest case of solving quadratic equations. If you look in some history of mathematics books (e.g., [**Hi:** Joseph] and [**Hi:** Eves]), you will find that quadratic equations were extensively solved by the Babylonians, Chinese, Indians, and Greeks. However, the earliest known general discussion of quadratic equations took place between 800 and 1100 A.D. in the Muslim Empire. Best known are Mohammed Ibn Musa al'Khowarizmi (who lived in Baghdad from 780 to 850 and from whose name we get our word "algorithm") and Omar Khayyam (the Persian geometer who is mostly known in the West for his philosophical poetry *The Rubaiyat*). Both wrote books whose titles contain the phrase *Al-jabr w'al mugabalah* (from which we get our word "algebra"), al'Khowarizmi in about 820 and Khayyam in about 1100. (See [**Hi:** Katz], Section 7.2.1.) An English translation of both books is available in many libraries, if you can figure out whose name it is catalogued under (see [**AT:** al'Khowarizmi] and [**AT:** Khayyam, 1931]). We previously met Khayyam in Chapter 12.

In these books you find geometric and numerical solutions to quadratic equations and geometric proofs of these solutions. But you will quickly notice that there is not one general quadratic equation as we are used to it:

$$ax^2 + bx + c = 0.$$

Rather, because the use of negative coefficients and negative roots was avoided, they list six types of quadratic equations (we follow Khayyam's lead and set the coefficient of x^2 equal to 1):

1. $bx = c$, which needs no solution,

2. $x^2 = bx$, which is easily solved,

3. $x^2 = c$, which has root $x = \sqrt{c}$,

4. $x^2 + bx = c$, with root $x = \sqrt{(b/2)^2 + c} - b/2$,

5. $x^2 + c = bx$, with roots $x = b/2 \pm \sqrt{(b/2)^2 - c}$, if $c < (b/2)^2$, and

6. $x^2 = bx + c$, with root $x = b/2 + \sqrt{(b/2)^2 + c}$.

Here b and c are always positive numbers or a geometric length (b) and area (c).

a. *Show that these are the only types. Why is $x^2 + bx + c = 0$ not included? Explain why b must be a length but c an area.*

The avoidance of negative numbers was widespread until a few hundred years ago. In the sixteenth century, European mathematicians called the negative numbers that appeared as roots of equations "numeri fictici" — fictitious numbers (see [**AT:** Cardano, page 11]). In 1759 Francis Masères (1731–1824, English), mathematician and a Fellow at Cambridge University and a member of the Royal Society, wrote in his *Dissertation on the Use of the Negative Sign in Algebra,*

> ...[negative roots] serve only, as far as I am able to judge, to puzzle the whole doctrine of equations, and to render obscure and mysterious things that are in their own nature exceeding plain and simple.... It were to be wished therefore that negative roots had never been admitted into algebra or were again discarded from it: for if this were done, there is good reason to imagine, the objections which many leaned and ingenious men now make to algebraic computations, as being obscure and perplexed with almost unintelligible notions, would be thereby removed; it being certain that Algebra, or universal arithmetic, is, in its own nature, a science no less simple, clear, and capable of demonstration, than geometry.

More recently in 1831, Augustus De Morgan (1806–1871, English), the first professor of mathematics at University College, London, and a founder of the London Mathematical Society, wrote in his *On the Study and Difficulties of Mathematics,*

> The imaginary expression $\sqrt{-a}$ and the negative expression $-b$ have this resemblance, that either of them occurring as solution of a problem indicates some inconsistency or absurdity. As far as real meaning is concerned, both are equally imaginary, since $0 - a$ is as inconceivable as $\sqrt{-a}$.

Why did these mathematicians avoid negative numbers and why did they say what they said? To get a feeling for why, think about the meaning of 2×3 as two 3's and 3×2 as three 2's and then try to find a meaning for $3 \times (-2)$ and $-2 \times (+3)$. Also consider the quotation at the beginning of this chapter from Omar Khayyam about algebra and geometry. Some historians have quoted this passage but have left out all the words appearing after "proved." In my opinion, this omission changes the meaning of the passage. Euclid's propositions that are mentioned by Khayyam are the basic ingredients of Euclid's proof of the square root construction and form a basis for the construction of conic sections — see Problem **18.2**, below. Geometric justification when there are negative coefficients is at the least very cumbersome, if not impossible. (If you doubt this, try to modify some of the geometric justifications below.)

    b. *Find geometrically the algebraic equations that express all the positive roots of each of the six types. Fill in the details in the following sketch of Khayyam's methods for Types 3–6.*

For the geometric justification of Type 3 and the finding of square roots, Khayyam refers to Euclid's construction of the square root in Proposition II 14, which we discussed in Problem **14.1**.

    For Type 4, Khayyam gives as geometric justification the illustration shown in Figure 18.1.

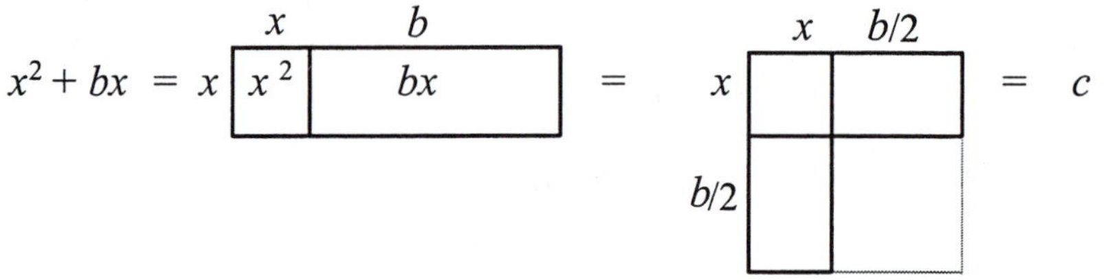

Figure 18.1   Type 4

Thus, by "completing the square" on $x + b/2$, we have

$$(x + b/2)^2 = c + (b/2)^2.$$

Thus we have $x = ...$? Note the similarity between this and Baudhayana's construction of the square root (see Chapter 13).

    For Type 5, Khayyam first assumes $x < (b/2)$ and draws the equation as Figure 18.2.

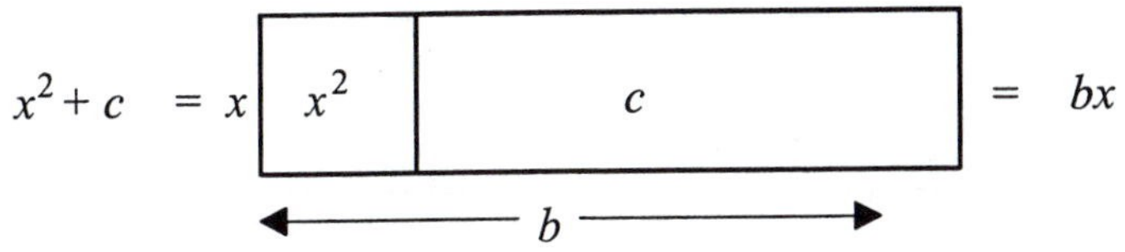

$$x^2 + c \;=\; x \;\boxed{\begin{array}{c} x^2 \end{array}\ \ c} \;=\; bx$$

Figure 18.2   Type 5, $x < (b/2)$

Note (see Figure 18.3) that the square on $b/2$ is $(b/2 - x)^2 + c$.

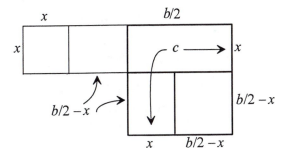

Figure 18.3   Type 5, $x < (b/2)$

This leads to $x = b/2 + \sqrt{(b/2)^2 + c}$. Note that if $c > (b/2)^2$, this geometric solution is impossible. When $x > (b/2)$, Khayyam uses the drawings shown in Figure 18.4.

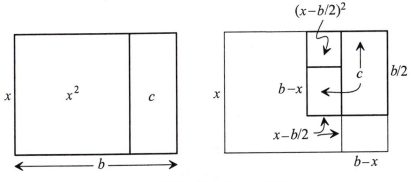

Figure 18.4   Type 5, $x > (b/2)$

For solutions of Type 6, Khayyam uses the drawing in Figure 18.5.

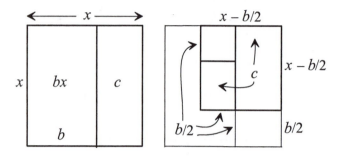

Figure 18.5   Type 6

Do the above solutions find the negative roots? The answer is clearly No if you mean, "Did al'Khowarizmi and Khayyam mention negative roots?" But let us not be too hasty. Suppose $-r$ (r, positive) is the negative root of $x^2 + bx = c$. Then $(-r)^2 + b(-r) = c$ or $r^2 = br + c$. Thus r is a positive root of $x^2 = bx + c$! The absolute value of the negative root of $x^2 + bx = c$ is the positive root of $x^2 = bx + c$ and vice versa. Also, the absolute values of the negative roots of $x^2 + bx + c = 0$ are the positive roots of $x^2 + c = bx$. So, in this sense, *Yes, the above geometric solutions do find all the real roots of all quadratic equations.* Thus it is misleading to state, as most historical accounts do, that the geometric methods failed to find the negative roots. The users of these methods did not find negative roots because they did not conceive of them. However, the methods can be directly used to find all the positive and negative roots of all quadratics.

    c. *Use Khayyam's methods to find all roots of the following equations: $x^2 + 2x = 2$, $x^2 = 2x + 2$, $x^2 + 3x + 1 = 0$.*

PROBLEM 18.2 CONIC SECTIONS AND CUBE ROOTS

The Greeks (for example, Archytas of Tarentum, 428–347 B.C., who was a Pythagorean in southern Italy, and Hippocrates of Chios in Asia Minor, 5th century B.C.) noticed that, if $a/c = c/d = d/b$, then $(a/c)^2 = (c/d)(d/b)$ $= (c/b)$ and, thus, $c^3 = a^2 b$. (For more historical discussion, see [**Hi:** van der Waerden], page 150, and [**Hi:** Katz], Chapter 2.) Now setting $a = 1$, we see that we can find the cube root of b, if we can find c and d such that $c^2 = d$ and $d^2 = bc$. If we think of c and d as being variables and b as a constant, then we see these equations as the equations of two parabolas with perpendicular axes and the same vertex. The Greeks also saw it this way but first they had to develop the concept of a parabola! The first general construction of conic sections was done by Menaechmus (a member of Plato's Academy) in the fourth century B.C.

To the Greeks, and later Khayyam, if AB is a line segment, then *the parabola with vertex B and parameter AB* is the curve P such that, if C is on P, then the rectangle $BDCE$ (see Figure 18.6) has the property that $(BE)^2 = BD \cdot AB$. Because in Cartesian coordinates the coordinates of C are (BE,BD), this last equation becomes a familiar equation for a parabola.

Points of the parabola may be constructed by using the construction for the square root given in Chapter 13. In particular, E is the intersection of the semicircle on AD with the line perpendicular to AB at B. (The construction can also be done by finding D' such that $AB = DD'$. The semicircle on BD' then intersects P at C.) I encourage you to try this construction yourself; it is very easy to do if you use a compass and graph paper.

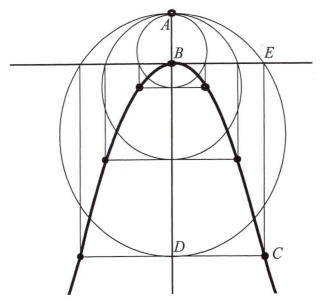

Figure 18.6   Construction of parabola

Now we can find the cube root. Let b be a positive number or length and let $AB = b$ and construct C so that CB is perpendicular to AB and such that $CB = 1$. See Figure 18.7. Construct a parabola with vertex B and parameter AB and construct another parabola with vertex B and parameter CB. Let E be the intersection of the two parabolas. Draw the rectangle $BGEF$. Then

$$(EF)^2 = BF{\cdot}AB \text{ and } (GE)^2 = GB{\cdot}CB.$$

But, setting $c = GE = BF$ and $d = GB = EF$, we have

$$d^2 = cb \text{ and } c^2 = d. \text{ Thus } c^3 = b.$$

If you use a fine graph paper, it is easy to get three-digit accuracy in this construction.

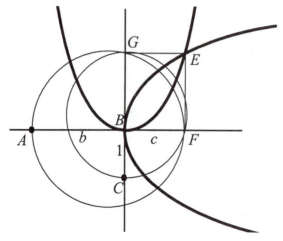

Figure 18.7   Finding cube roots

The Greeks did a thorough study of conic sections and their properties, culminating in Apollonius's (c. 260–170 B.C., Greek) book *Conics*, which appeared around 200 B.C. You can read this book in English translation (see [**AT:** Apollonius]).

    a. *Use the above geometric methods with a fine graph paper to find the cube root of 10.*

To find roots of cubic equations we shall also need to know *the (rectangular) hyperbola with vertex B and parameter AB.* This is the curve such that, if E is on the curve and $ACED$ is the determined rectangle (see Figure 18.8), then $(EC)^2 = BC \cdot AC$.

The point E can be determined using the construction from Chapter 13. Let F be the bisector of AB. Then the circle with center F and radius FC will intersect at D the line perpendicular to AB at A. From the drawing it is clear how these circles also construct the other branch of the hyperbola (with vertex A.)

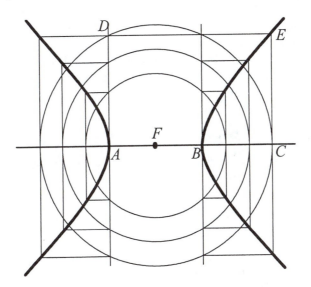

Figure 18.8. Construction of hyperbola

b. *Use the above method with graph paper to construct the graph of the hyperbola with parameter 5. What is an algebraic equation that represents this hyperbola?*

Notice how these descriptions and constructions of the parabola and hyperbola look very much as if they were done in Cartesian coordinates. *The ancestral forms of Cartesian coordinates and analytic geometry are evident here.* Also they are evident in the solutions of cubic equations in the next section. The ideas of Cartesian coordinates did not appear to Rene Descartes (1596–1650, French) out of nowhere. The underlying concepts were developing in Greek and Muslim mathematics. One of the apparent reasons that full development did not occur until Descartes is that, as we have seen, negative numbers were not accepted. The full use of negative numbers is essential for the realization of Cartesian coordinates. However, even Descartes seems to have avoided negatives as much as possible when he was studying curves — he would start with a curve (constructed by some geometric or mechanical procedure) and then choose axes so that the important parts of the curve had both coordinates positive. However, it is not true (as asserted in a well-known history of mathematics book) that Descartes always used x and y to stand for positive values. For example, in Book II of his *Geometrie* [**AT:** Descartes] he is describing the construction of a locus generated by a point C and defines [on page 60] $y = \mathbf{CB}$, where B is a given point, and derives an equation satisfied by y and other variables; and then, in the same paragraph [page 63], continues "If y is zero or less than nothing in this equation ..."

PROBLEM **18.3** ROOTS OF CUBIC EQUATIONS

In his *Al-jabr wa'l muqabalah,* Omar Khayyam also gave geometric solutions to cubic equations. You will see that his methods are sufficient to find geometrically all real (positive or negative) roots of cubic equations; however; in his first chapter Khayyam says (see [**AT:** Khayyam (1931)], page 49)

> When, however, the object of the problem is an absolute number, neither we, nor any of those who are concerned with algebra, have been able to prove this equation — perhaps others who follow us will be able to fill the gap — except when it contains only the three first degrees, namely, the number, the thing and the square.

By "absolute number," Khayyam is referring to what we call algebraic solutions, as opposed to geometric ones. This quotation suggests, contrary to what many historical accounts say, that Khayyam expected that algebraic solutions would be found.

Khayyam found 19 types of cubic equations (when expressed with only positive coefficients). (See [**AT:** Khayyam (1931)], page 51.) Of these 19, 5 reduce to quadratic equations (for example, $x^3 + ax = bx$ reduces to $x^2 + ax = b$). The remaining 14 types Khayyam solves by using conic sections. His methods find all the positive roots of each type, although he failed to mention some of the roots in a few cases, and, of course, he ignores the negative roots. Instead of going through his 14 types, I will show how a simple reduction will reduce them to only four types in addition to types already solved, such as $x^3 = b$. I will then give Khayyam's solutions to these four types.

In the cubic $y^3 + py^2 + gy + r = 0$ (where p, g, r, are positive, negative, or zero), set $y = x - (p/3)$. Try it! The resulting equation in x will have the form $x^3 + sx + t = 0$, (where s and t are positive, negative, or zero). If we rearrange this equation so all the coefficients are positive, we get the following four types that have not been previously solved:

$$(1) \ x^3 + ax = b, \ (2) \ x^3 + b = ax,$$

$$(3) \ x^3 = ax + b, \ \text{and} \ (4) \ x^3 + ax + b = 0,$$

where a and b are positive, in addition to the types previously solved.

  a.   *Show that in order to find all the roots of all cubic equations we need only have a method that finds the roots of Types 1, 2, and 3.*

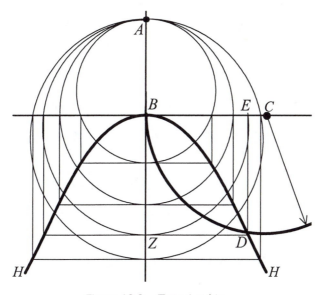

Figure 18.9   Type 1 cubic

KHAYYAM'S SOLUTION FOR TYPE 1: $x^3 + ax = b$

A cube and sides are equal to a number. Let the line AB [see Figure 18.9] be the side of a square equal to the given number of roots [that is, $(AB)^2 = a$, the coefficient]. Construct a solid whose base is equal to the square on AB, equal in volume to the given number [b]. The construction has been shown previously. Let BC be the height of the solid. [That is, $BC \cdot (AB)^2 = b$.] Let BC be perpendicular to AB ... Construct a parabola whose vertex is the point B ... and parameter AB. Then the position of the conic HBD will be tangent to BC. Describe on BC a semicircle. It necessarily intersects the conic. Let the point of intersection be D; drop from D, whose position is known, two perpendiculars DZ and DE on BZ and BC. Both the position and magnitude of these lines are known.

The root is EB. Khayyam's proof (using a more modern, compact notation) is as follows: From the properties of the parabola (Problem **18.2**) and circle (Problem **14.1**) we have

$$(DZ)^2 = (EB)^2 = BZ \cdot AB \ \text{ and } \ (ED)^2 = (BZ)^2 = EC \cdot EB,$$

thus

$$EB \cdot (BZ)^2 = (EB)^2 \cdot EC = BZ \cdot AB \cdot EC$$

and, therefore,

$$AB \cdot EC = EB \cdot BZ,$$

and

$$(EB)^3 = EB \cdot (BZ \cdot AB) = (AB \cdot EC) \cdot AB = (AB)^2 \cdot EC.$$

So

$$(EB)^3 + a(EB) = (AB)^2 \cdot EC + (AB)^2 \cdot (EB) = (AB)^2 \cdot CB = b.$$

Thus EB is a root of $x^3 + ax = b$. Because $x^2 + ax$ increases as x increases, there can be only this one root.

KHAYYAM'S SOLUTIONS FOR TYPES 2 AND 3: $x^3 + b = ax$ AND $x^3 = ax + b$

Khayyam treated these equations separately but by allowing negative horizontal lengths we can combine his two solutions into one solution of $x^3 \pm b = ax$. Let AB be perpendicular to BC and as before let $(AB)^2 = a$ and $(AB)^2 \cdot BC = b$. Place BC to the left if the sign in front of b is negative

(Type 3) and place *BC* to the right if the sign in front of *b* is positive (Type 2). Construct a parabola with vertex *B* and parameter *AB*. Construct both branches of the hyperbola with vertices *B* and *C* and parameter *BC*. See Figure 18.10.

Each intersection of the hyperbola and the parabola (except for *B*) gives a root of the cubic. Suppose they meet at *D*. Then drop perpendiculars *DE* and *DZ*. The root is *BE* (negative if to the left and positive if to the right). Again, if you use fine graph paper, it is possible to get three-digit accuracy here. I leave it for you, the reader, to provide the proof, which is very similar to Type 1.

  b. *Verify that Khayyam's method described above works for Types 2 and 3. Can you see from your verification why the extraneous root given by B appears?*

  c. *Use Khayyam's method to find all solutions to the cubic*

$$x^3 = 15x + 4.$$

Use fine graph paper and try for three-place accuracy.

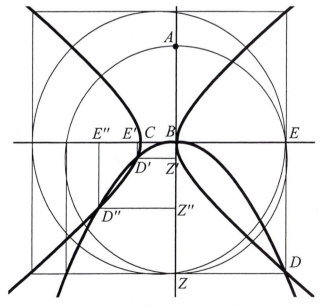

Figure 18.10   Types 2 and 3 cubics

PROBLEM 18.4   ALGEBRAIC SOLUTION OF CUBICS

A little more history: Most historical accounts assert correctly that Khayyam did not find the negative roots of cubics. However, they are misleading in that they all fail to mention that his methods are fully sufficient to find the negative roots, as we have seen above. This is in contrast to the common assertion (see, for example, [**SE:** Davis & Hersch]) that Girolamo Cardano (1501–1578, Italian) was the first to publish the general solution of cubic equations. In fact, as we shall see, Cardano himself admitted that his methods are insufficient to find the real roots of many cubics.

Cardano published his algebraic solutions in his book, *Artis Magnae* (The Great Art), which was published in 1545. For a readable English translation and historical summary, see [**AT:** Cardano]. Cardano used only positive coefficients and thus divided the cubic equations into the same 13 types (excluding $x^3 = c$ and equations reducible to quadratics) used earlier by Khayyam. Cardano also used geometry to prove his solutions for each type. As we did above, we can make a substitution to reduce these to the same types as above:

$$(1)\ x^3 + ax = b,\ (2)\ x^3 + b = ax,$$

$$(3)\ x^3 = ax + b,\ \text{and}\ (4)\ x^3 + ax + b = 0.$$

If we allow ourselves the convenience of using negative numbers and lengths, then we can reduce these to one type: $x^3 + ax + b = 0$, where now we allow a and b to be either negative or positive.

The main "trick" that Cardano used was to assume that there is a solution of $x^3 + ax + b = 0$ of the form $x = t^{1/3} + u^{1/3}$. Plugging this into the cubic we get

$$(t^{1/3} + u^{1/3})^3 + a(t^{1/3} + u^{1/3}) + b = 0.$$

If you expand and simplify this, you get to

$$t + u + b + (3t^{1/3}u^{1/3} + a)(t^{1/3} + u^{1/3}) = 0.$$

(Cardano did this expansion and simplification geometrically by imagining a cube with sides $t^{1/3} + u^{1/3}$.) Thus $x = t^{1/3} + u^{1/3}$ is a root if

$$t + u = -b\ \text{and}\ t^{1/3}\,u^{1/3} = -(a/3).$$

Solving, we find that t and u are the roots of the quadratic equation

$$z^2 + bz - (a/3)^3 = 0$$

which Cardano solved geometrically (and so can you, Problem **18.1**) to get

$$t = -b/2 + \sqrt{(b/2)^2 + (a/3)^3} \quad \text{and} \quad u = -b/2 - \sqrt{(b/2)^2 + (a/3)^3} \ .$$

Thus the cubic has roots

$$x = t^{1/3} + u^{1/3}$$

$$= \{-b/2 + \sqrt{(b/2)^2 + (a/3)^3}\ \}^{1/3} + \{-b/2 - \sqrt{(b/2)^2 + (a/3)^3}\ \}^{1/3}.$$

This is Cardano's cubic formula. But, a strange thing happened. Cardano noticed that the cubic $x^3 = 15x + 4$ has a positive real root 4 but, for this equation, $a = -15$ and $b = -4$, and if we put these values into his cubic formula, we get that the roots of $x^3 = 15x + 4$ are

$$x = \{2 + \sqrt{-121}\}^{1/3} + \{2 - \sqrt{-121}\}^{1/3}\ .$$

But these are the sum of two complex numbers even though you have shown in Problem **18.3** that all three roots are real. How can this expression yield 4?

In Cardano's time there was no theory of complex numbers and so he reasonably concluded that his method would not work for this equation, even though he did investigate expressions such as $\sqrt{-121}$. Cardano writes ([**AT:** Cardano, page 103])

> When the cube of one-third the coefficient of x is greater than the square of one-half the constant of the equation ... then the solution of this can be found by the aliza problem which is discussed in the book of geometrical problems.

It is not clear what book he is referring to, but the "aliza problem" presumably refers to the mathematician known as al'Hazen, Abu Ali al'Hasan ibu al'Haitam (965–1039), who was born in Persia and worked in Egypt and whose works were known in Europe in Cardano's time. Al'Hazen had used intersecting conics to solve specific cubic equations and the problem of describing the image seen in a spherical mirror — this latter problem in some books is called "Alhazen's problem."

In addition, we know today that each complex number has three cube roots and so the formula

$$x = \{2 + \sqrt{-121}\}^{1/3} + \{2 - \sqrt{-121}\}^{1/3}$$

is ambiguous. In fact, some choices for the two cube roots give roots of the cubic and some do not. (Experiment with $x^3 = 15x + 4$.) Faced with Cardano's Formula and equations such as $x^3 = 15x + 4$, Cardano and other mathematicians of the time started exploring the possible meanings of these complex numbers and thus started the theory of complex numbers.

a. *Solve the cubic $x^3 = 15x + 4$ using Cardano's Formula and your knowledge of complex numbers.*

Remember that on the previous page we showed that $x = t^{1/3} + u^{1/3}$ is a root of the equation if $t + u = -b$ and $t^{1/3} u^{1/3} = -(a/3)$.

b. *Solve $x^3 = 15x + 4$ by dividing through by $x - 4$ and then solving the resulting quadratic.*

c. *Compare your answers and methods of solution from Problems 18.3c, 18.4a, and 18.4b.*

SO WHAT DOES THIS ALL POINT TO?

So what does the experience of this chapter point to? It points to different things for each of us. I conclude that it is worthwhile paying attention to the meaning in mathematics. Often in our haste to get to the modern, powerful analytic tools we ignore and trod upon the meanings and images that are there. Sometimes it is hard even to get a glimpse that some meaning is missing. One way to get this glimpse and find meaning is to listen to and follow questions of "What does it mean?" that come up in ourselves, in our friends, and in our students. We must listen creatively because we and others often do not know how to express precisely what is bothering us.

Another way to find meaning is to read the mathematics of old and keep asking, "Why did they do that?" or "Why didn't they do this?" Why did the early algebraists (up until at least 1600 and much later, I think) insist on geometric proofs? I have suggested some reasons above.

Today, we normally pass over geometric proofs in favor of analytic ones based on the 150-year-old notion of Cauchy sequences and the Axiom of Completeness. However, for most students and, I think, most mathematicians, our intuitive understanding of the real numbers is based on the geometric real line. As an example, think about multiplication: What does $a \times b$ mean? Compare the geometric images of $a \times b$ with the multiplication of two infinite, non-repeating, decimal fractions. What is $\sqrt{2} \times \pi$?

There is another reason why a geometric solution may be more meaningful: Sometimes we actually desire a geometric result instead of a numerical one. For example, a friend and I were building a small house using wood. The roof of the house consisted of 12 isosceles triangles which together formed a 12-sided cone (or pyramid). It was necessary for us to determine the angle between two adjacent triangles in the roof so we could appropriately cut the log rafters. I immediately started to calculate the angle using (numerical) trigonometry and algebra. But then I ran into a problem. I had only a slide rule with three-place accuracy for finding square roots and values of trigonometric functions. At one point in the calculation I had to subtract two numbers that differed only in the third place (for example, $5.68 - 5.65$); thus my result had little accuracy. As I started to figure out a different computational procedure that would avoid the subtraction, I suddenly realized *I didn't want a number, I wanted a physical angle*. In fact, a numerical angle would be essentially useless — imagine taking two rough boards and putting them at a given numerical angle apart using only an ordinary protractor! What I needed was the physical angle, full-size. So I constructed the angle on the floor of the house using a rope as a compass. This geometric solution had the following advantages over a numerical solution:

- The geometric solution resulted in the desired physical angle, while the numerical solution resulted in a number.

- The geometric solution was quicker than the numerical solution.

- The geometric solution was immediately understood and trusted by my friend (and fellow builder), who had almost no mathematical training, while the numerical solution was

beyond my friend's understanding because it involved trigonometry (such as the "Law of Cosines").

◆ And, because the construction was done full-size, the solution automatically had the degree of accuracy appropriate for
the application.

**Meaning is important in mathematics, and geometry is an important
source of that meaning.**

Chapter 19

TRIGONOMETRY AND DUALITY

After we have found the equations [The Laws of Cosines and Sines for a Hyperbolic plane] which represent the dependence of the angles and sides of a triangle; when, finally, we have given general expressions for elements of lines, areas and volumes of solids, all else in the [Hyperbolic] Geometry is a matter of analytics, where calculations must necessarily agree with each other, and we cannot discover anything new that is not included in these first equations from which must be taken all relations of geometric magnitudes, one to another. ... We note however, that these equations become equations of spherical Trigonometry as soon as, instead of the sides a, b, c we put $a\sqrt{-1}$, $b\sqrt{-1}$, $c\sqrt{-1}$ ...

— N. Lobachevsky, quoted in [**NE:** Greenberg]

In this chapter, we will first derive, geometrically, expressions for the circumference of a circle on a sphere, the Law of Cosines on the plane, and its analog on a sphere. Then we will talk about duality on a sphere. On a sphere, duality will enable us to derive other laws that will help our two-dimensional bug to compute sides and/or angles of a triangle given ASA, RLH, SSS, or AAA. Finally we will look at duality on the plane.

PROBLEM 19.1 CIRCUMFERENCE OF A CIRCLE

a. *Find a simple formula for the circumference of a circle on a sphere in terms of its intrinsic radius and make the formula as intrinsic as possible.*

We suggest that you make an extrinsic drawing (similar to Figure 19.1) of the circle, its intrinsic radius, its extrinsic radius, and the center of the sphere. You may well find it convenient to use trigonometric functions to express your answer. Note that the existence of trigonometric functions for right triangles follows from the properties of similar triangles that were proved in Problem **13.3**.

In Figure 19.1, rotating the segment of length r' (***the extrinsic radius***) through a whole revolution produces the same circumference as rotating r, which is an arc of the great circle as well as ***the intrinsic radius*** of the circle on the sphere.

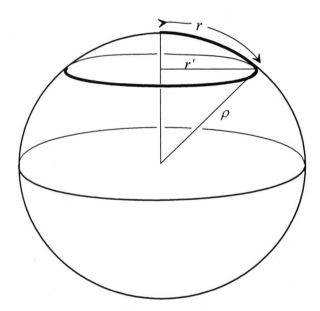

Figure 19.1   Intrinsic radius r

Even though the derivation of the formula this way will be extrinsic, it is possible, in the end, to express the circumference only in terms of intrinsic quantities. Thus, also think of the following problem:

b. *How could our 2-dimensional bug derive this formula?*

By looking at very small circles, the bug could certainly find uses for the trigonometric functions they give rise to. Then the bug could discover that the geodesics are actually (intrinsic) circles, but circles that do not have the same trigonometric properties as very small circles. And then what?

Using the expressions of trigonometric and hyperbolic functions in terms of infinite series, it is proved (in [**NE:** Greenberg], page 337) that

Theorem 19.1. *In a hyperbolic plane of radius 1, a circle with intrinsic radius r has circumference c equal to*

$$c = 2\pi \sinh(r).$$

c. *Use the theorem to show that on a hyperbolic plane of radius ρ, a circle with intrinsic radius r has circumference c equal to*

$$c = 2\pi\, \rho \sinh(r/\rho).$$

When going from a hyperbolic plane of radius 1 to a hyperbolic plane of radius ρ, all lengths scale by a factor of ρ. *Why?*

The formula in Part **c** should look very much like your formula for Part **a** (possibly with some algebraic manipulations). This is precisely what Lobachevsky was talking about in the quote at the beginning of this chapter. Check this out with Part **d**.

d. *Show that, if you replace ρ by $i\rho$ in the formula of Part **c**, then you will get the formula in Part **a**.*

Look up the definition of *sinh* (hyperbolic sine) and express it as a Taylor series.

Problem **19.2**   Law of Cosines

If we know two sides and the included angle of a (small) triangle, then according to SAS the third side is determined. If we know the lengths of the two sides and the measure of the included angle, how can we find the length of the third side? The various formulas that give this length are called the ***Law of Cosines***.

a. *Find a law of cosines for triangles in the plane.*

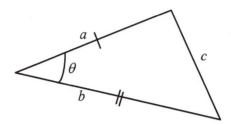

Figure 19.2   Law of Cosines

You have learned in school (but perhaps forgotten) the **Law of Cosines** on the plane: $c^2 = a^2 + b^2 - 2ab \cos \theta$. For a geometric proof of this "law," look at the pictures in Figure 19.3. I first saw the idea for these pictures in the marvelous book [**Hi:** Valens]. These pictures show the squares as rigid with hinges at all the points marked ⌒. Note that in the middle picture θ is greater than $\pi/2$. You must draw a different picture for θ less than $\pi/2$. Prove the Law of Cosines on the plane using the pictures in Figure 19.3, or in any other way you wish.

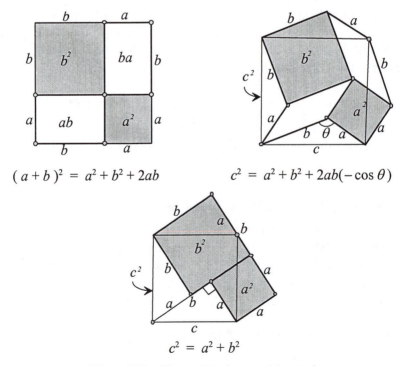

$$(a + b)^2 = a^2 + b^2 + 2ab$$

$$c^2 = a^2 + b^2 + 2ab(-\cos \theta)$$

$$c^2 = a^2 + b^2$$

Figure 19.3   Three related geometric proofs

b. *Find a law of cosines for small triangles on a sphere with radius* ρ.

On the sphere there are various versions of the Law of Cosines that you can find. One approach that will work is to project the triangle by a gnomic projection onto the plane tangent to the sphere at the vertex of the given angle. This projection will preserve the size of the given angle (*Why?*) and, even though it will not preserve the lengths of the sides of the triangle, you can determine what effect it has on these lengths. Now apply the planar Law of Cosines to this projected triangle. It is very helpful to draw a 3-D picture of this projection.

Pause, explore, and write about this problem before you read further.

It is sometimes convenient to measure lengths of great circle arcs on the sphere in terms of the radian measure. In particular,

radian measure of the arc = (length of the arc)/ρ,

where ρ is the radius of the sphere. For example, the radian measure of one quarter of a great circle would be $\pi/2$ and the radian measure of half a great circle would be π. In Figure 19.4, the segment a is subtended by the angle α at the pole and by the same angle α at the center of the sphere. The radian measure of a is the radian measure of α.

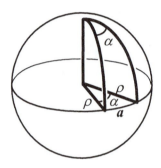

Figure 19.4   Radian measure of lengths

If we measure lengths in radians, then one possible formula for the spherical triangle of radius ρ in Part **b** is:

$$\cos c = \cos a \cos b + \sin a \sin b \cos \theta.$$

(This is not the only such formula.)

For a right triangle ($\theta = \pi/2$) the above formula becomes

$$\cos c/\rho = \cos a/\rho \cos b/\rho,$$

which can be considered as the spherical equivalent of the Pythagorean Theorem.

THEOREM 19.2. *A Law of Cosines for triangles on a hyperbolic plane with radius ρ is*

$$\cosh c/\rho = \cosh a/\rho \cosh b/\rho + \sinh a/\rho \sinh b/\rho \cos \theta.$$

This theorem is proved in [**NE:** Stahl], page 125, using analytic techniques and in [**NE:** Greenberg] using infinite series representations.

This is one of the "equations" that Lobachevsky is talking about in the quote at the beginning of the chapter. It is important to realize that a study of hyperbolic trigonometric functions was started by V. Riccati (1707–1775, Italy) and continued by J.H. Lambert (1728–1777, Germany) in 1768 well before their use in hyperbolic geometry by Lobachevsky.

PROBLEM **19.3** LAW OF SINES

Closely related to the Law of Cosines is the *Law of Sines*.

    a. *If △ABC is a triangle on the plane with sides, a, b, c, and corresponding opposite angles, α, β, γ, then*

$$a/\sin\alpha = b/\sin\beta = c/\sin\gamma.$$

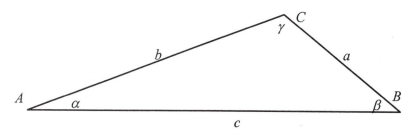

Figure 19.5   Law of Sines

    The standard proof for the Law of Sines is to drop a perpendicular from the vertex C to the side c and then to express the length of this perpendicular as both ($b \sin \alpha$) and ($a \sin \beta$). See Figure 19.6. From this the result easily follows. Thus, on the plane the Law of Sines follows from an expression for the sine of an angle in a right triangle.

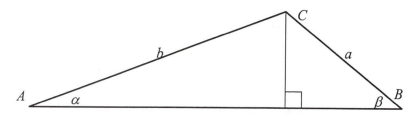

Figure 19.6   Standard proof of Law of Sines on plane

For triangles on the sphere we can find a very similar result.

b. *What is an analogous property on the sphere?*

If $\triangle ACD$ is a triangle on the sphere with the angle at D being a right angle, then use gnomic projection to project $\triangle ACD$ onto the plane that is tangent to the sphere at A. Because the plane is tangent to the sphere at A, the size of the angle α is preserved under the projection. In general, angles on the sphere not at A will not be projected to angles of the same size, but in this case the right angle at D will be projected to a right angle. (*Be sure you see why this is the case. A good drawing and the use of symmetry and similar triangles will help.*) Now express the sine of α in terms of the sides of this projected triangle.

When we measure the sides in radians, on the sphere the Law of Sines becomes

$$(\sin a)/\sin \alpha = (\sin b)/\sin \beta = (\sin c)/\sin \gamma.$$

For a right triangle this becomes

$$\sin \alpha = (\sin a)/(\sin c).$$

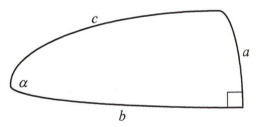

Figure 19.7   Law of Sines for right triangles on a sphere

DUALITY ON A SPHERE

We can now ask: Is it possible to find expressions corresponding to the other triangle congruence theorems which we proved in Chapters 6 and 9. Let us see how we can be helped by a certain concept of duality which we will now develop.

When we were looking at SAS and ASA, we noticed a certain duality between points and lines (geodesics). SAS was true on the plane or open hemisphere because *two points determine a unique line segment* and ASA was true on the plane or open hemisphere because *two (inter-secting) lines determine a unique point*. In this section we will make this notion of duality broader and deeper and look at it in such a way that it applies to both the plane and the sphere.

On the whole sphere, two distinct points determine a unique straight line (great circle) *unless the points are antipodal*. In addition, two distinct great circles determine a unique *pair of antipodal points*. Also, a circle on the sphere has two centers *which are antipodal*. Remember also that in most of the triangle congruence theorems, we had *trouble with triangles that contained antipodal points*. So our first step is to consider not points on the sphere but rather ***point-pairs***, pairs of antipodal points. With this definition in mind, check the following:

- *Two distinct point-pairs determine a unique great circle (geodesic).*

- *Two distinct great circles determine a unique point-pair.*

- *The intrinsic center of a circle is a single point-pair.*

- *SAS, ASA, SSS, AAA are true for all triangles not containing any point-pairs.*

Now, we can make the duality more definite.

- *The dual of a great circle is its **poles*** (the point-pair that is the intrinsic center of the great circle). Some books use the term "***polar***" in place of "***dual***".

- *The dual of a point-pair is its **equator*** (the great circle whose center is the point-pair).

Notice: *If the point-pair **P** is on the great circle **l**, then the dual of **l** is on the dual of **P**. See Figure 19.8.

If k is another great circle through P, then notice that the dual of k is also on the dual of P. Because an angle can be viewed as a collection of lines (great circles) emanating from a point, the dual of this angle is a collection of point-pairs lying on the dual of the angle's vertex. And, vice versa, the dual of the points on a segment of a great circle are great circles emanating from a point-pair that is the dual of the original great circle. *Before going on be sure to understand this relationship between an angle and its dual.* Draw pictures. Make models.

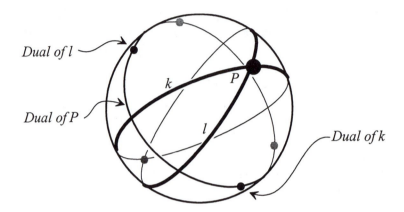

Dual of l

k

P

Dual of P

l

Dual of k

Figure 19.8   *P* is on *l* implies the dual of *l* is on the dual of *P*

PROBLEM 19.4   THE DUAL OF A SMALL TRIANGLE

The dual of the small triangle △*ABC* is the small triangle △*A*B*C**, where *A** is that pole of the great circle of *BC* which is on the same side of *BC* as the vertex *A*, similarly for *B**, *C**. See Figure 19.9.

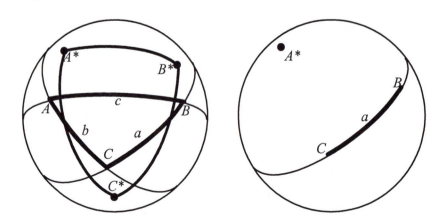

Figure 19.9   The dual of a small triangle

a. *Find the relationship between the sizes of the angles and sides of a triangle and the corresponding sides and angles of its dual.*

b. *Is there a triangle that is its own dual?*

*PROBLEM **19.5** TRIGONOMETRY WITH CONGRUENCES

a. *Find the dual of the Law of Cosines on the sphere.*

There is an analogous dual Law of Cosines for a hyperbolic plane, proved in the same books cited with Theorem 19.2, which proved the hyperbolic Law of Cosines.

b. *For each of ASA, RLH, SSS, AAA, if you know the measures of the given sides and angles, how can you find the measures of the sides and angles that are not given? Do this for both spheres and hyperbolic planes.*

Use Part **a** and the formulas from Problems **19.1** and **19.2**.

DUALITY ON THE PROJECTIVE PLANE

The gnomic projection, *g* (Problem **16.1**), allows us to transfer the above duality on the sphere to a duality on the plane. If *P* is a point on the plane, then there is a point *Q* on the sphere such that *g(Q) = P*. The dual of *Q* is a great circle *l* on the sphere. If *Q* is not the South Pole, then half of *l* is in the Southern Hemisphere and its projection onto the plane, *g(l)*, is a line we can call the ***dual of P***. This defines a dual for every point on the plane except for the point where the South Pole of the sphere rests. See Figure 19.10. It is convenient to call this point the origin, *O*, of the plane.

Note that *O* is the image of *S*, the South Pole, and that the dual of *S* is the equator, which is projected by *G* to infinity on the plane. Thus we define the dual of *O* to be the ***line at infinity***. If *l* is any line in the plane, then it is the image of a great circle on the sphere that intersects the equator in a point-pair. The image of this point-pair is considered to be a *single* point at infinity at the "end" of the line *l*, the *same* point at both ends. The plane with the line at infinity attached is called the ***projective plane***. This projective plane is the same as the *real projective plane* that we investigated in Problem **17.4b**, the only difference being that here we are focusing on the gnomic projection onto the plane and in Chapter 17 we were considering it as a spherical 2-manifold.

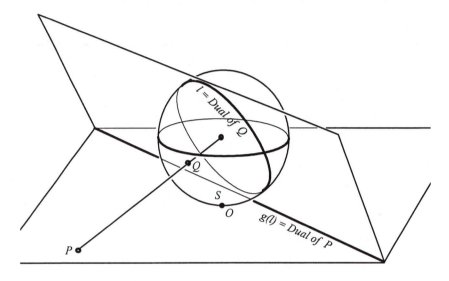

Figure 19.10   Duality on the projective plane

*PROBLEM **19.6** PROPERTIES ON THE PROJECTIVE PLANE

a. *Check that the following hold on the projective plane*:

♦ *Two points determine a unique line.*

♦ *Two parallel lines share the same point at infinity.*

♦ *Two lines determine a unique point.*

♦ *If a point is on a line, then the dual of the line is a point which is on the dual of the original point.*

b. *If γ is the circle with center at the origin and with radius the same as the radius of the sphere and if (P,Q) is an inversive pair with respect to γ, then show that the dual of P is the line perpendicular to OP at the point −Q, the point in the plane that is opposite (with respect to O) of Q. See Figure 19.11.*

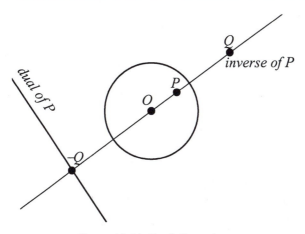

Figure 19.11 Duals/inversions

c. *Show that the dual of a point P on γ is a line tangent to γ at the diametrically opposite point –P.*

Dual Views of Our Experience

Our usual point of view for viewing our world is from the origin (because in this view we are the center of our world). We look out in all directions to observe what is outside and we strive to look out as close toward infinity as we can. The dual of this is the point of view where we are the line at infinity (the dual of the origin) and we view the whole world (which is within us) and we strain to look in as close toward the center as we can.

Perspective Drawings and Vision

Look at the perspective drawing in Figure 19.12.

Present-day theories of projective geometry got their start from Euclid's *Optics* [**AT**: Euclid] and later from the theories of perspective developed during the Renaissance by artists who studied the geometry inherent in perspective drawings. The "vanishing point" where the lines that are parallel on the house intersect on the horizon of the drawing is an image on the drawing of the point at infinity on these parallel lines.

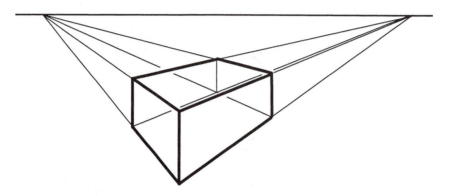

Figure 19.12   Perspective drawing

One way to visualize this is to imagine yourself at the center of a transparent sphere looking out at the world. If you look at two parallel straight lines and trace these lines on the sphere, you will be tracing segments of two great circles. If you followed this tracing indefinitely to the ends of the straight lines, you would have a tracing of two half great circles on the sphere intersecting in their endpoints. These endpoints of the semicircles are the images of the point at infinity that is the common intersection of the two parallel lines. If you now use a gnomic projection to project this onto a plane (for example, the artist's canvas), then you will obtain two straight line segments intersecting at one of their endpoints as in the drawing above.

Chapter 20

3-Spheres and Hyperbolic 3-Spaces

> Let us, then, make a mental picture of our universe: ...as far as possible, a complete unity so that whatever comes into view, say the outer orb of the heavens, shall bring immediately with it the vision, on the one plane, of the sun and of all the stars with earth and sea and all living things as if exhibited upon a transparent globe.
>
> Bring this vision actually before your sight, so that there shall be in your mind the gleaming representation of a sphere, a picture holding all the things of the universe... Keep this sphere before you, and from it imagine another, a sphere stripped of magnitude and of spatial differences; cast out your inborn sense of Matter, taking care not merely to attenuate it: call on God, maker of the sphere whose image you now hold, and pray Him to enter. And may He come bringing His own Universe...
>
> — Plotinus, *The Enneads*, **V.8.9**, Burdette, NY: Larson, 1992

In this chapter you will explore hyperbolic 3-space and the 3-dimensional sphere that extrinsically sits in 4-space. But intrinsically, if we zoom in on a point in a 3-sphere or a hyperbolic 3-space, then locally the experience of the space will become indistinguishable from an intrinsic and local experience of Euclidean 3-space. This is also our human experience in our physical universe. We will study these 3-dimensional spaces both because they are possible geometries for our physical universe, and in order to see that these geometries are closely related to their 2-dimensional versions.

The starred problems will need some experience with linear algebra, but can be skipped as long as you experience Problem **20.1** and at least assume the results of Problem **20.6**.

Try to imagine the possibility of our physical universe being a 3-sphere in 4-space. It is the same kind of imagination a 2-dimensional being would need in order to imagine that it was on a plane or 2-sphere (ordinary sphere) in 3-space.

PROBLEM 20.1   EXPLAIN 3-SPACE TO 2-D PERSON

How would you explain 3-space to a person living in two dimensions?

Think about the question in terms of this example: The person depicted in Figure 20.1 *lives in a 2-dimensional plane. The person is wearing a mitten on the right hand. Notice that there is no front or back side to the mitten for the 2-D person. The mitten is just a thick line around the hand.*

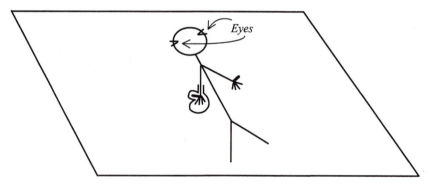

Figure 20.1   2-dimensional person with mitten

Suppose that you approach the plane, remove the mitten, and put it on the 2-D person's left hand. There's no way within 2-space to move the mitten to fit the other hand. So, you take the mitten off of the 2-D plane, flip it over in 3-space, and then put it back on the plane around the left hand. The 2-D person has no experience of three dimensions but can see the result — the mitten disappears from the right hand, the mitten is gone for a moment, and then it is on the left hand.

Figure 20.2   Where did the mitten go?

How would you explain to the 2-D person what happened to the mitten?

SUGGESTIONS

This person's 2-dimensional experience is very much like the experience of an insect called a water strider that we talked about in Chapter 2. A water strider walks on the surface of a pond and has a very 2-dimensional perception of the universe around it. To the water strider, there is no up or down; its whole universe consists of the surface of the water. Similarly, for the 2-D person there is no front or back; the entire universe is the 2-dimensional plane.

Living in a 2-D world, the 2-D person can easily understand any figures in 2-space, including planes. In order to explain a notion such as "perpendicular," we could ask the 2-D person to think about the thumb and fingers on one hand.

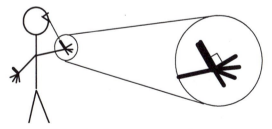

Figure 20.3   The 2-D person sees "perpendicular"

A person living in a 2-D world cannot directly experience three dimensions, just as we are unable to directly experience four dimensions. Yet, with some help from you, the 2-D person can begin to imagine three dimensions just as we can imagine four dimensions. One goal of this problem is to try to gain a better understanding of what our experience of 4-space might be. Think about what four dimensions might be like, and you may have ideas about the kinds of questions the 2-D person will have about three dimensions. You may know some answers, as well. The problem is finding a way to talk about them. Be creative!

One important thing to keep in mind is that it is possible to have *images* of things we cannot see. For example, when we look at a sphere, we can see only roughly half of it, but we can and do have an image of the entire sphere in our minds. We even have an image of the inside of the sphere, but it is impossible to actually see the entire inside or outside of the sphere all at once. Another example — sit in your room, close your eyes, and try to imagine the entire room. It is likely that you will have an image of the entire room, even though you can never see it all at once. Without such images of the whole room it would be difficult to maneuver around the room. The same goes for your image of the whole of the chair you are sitting on or this book you are reading.

Assume that the 2-D person also has images of things that cannot be seen in their entirety. For example, the 2-D person may have an image of a circle. Within a 2-dimensional world, the entire circle cannot be seen all at once; the 2-D person can only see approximately half of the outside of the circle at a time and cannot see the inside at all unless the circle is broken.

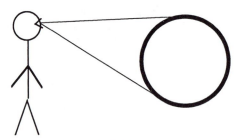

Figure 20.4   The 2-D person sees a circle

However, from our position in 3-space we *can* see the entire circle including its inside. Carrying the distinction between what we can see and what we can imagine one step further, the 2-D person cannot see the entire circle but can imagine in the mind the whole circle including inside and out. Thus, the 2-D person can only imagine what we, from three dimensions, can directly see. So, the 2-D person's image of the entire circle is as if it were being viewed from the third dimension. It makes sense, then, that the image of the entire sphere that we have in our minds is a 4-D view of it, as if we were viewing it from the fourth dimension.

When we talk about the fourth dimension here, we are not talking about time, which is often considered the fourth dimension. Here, we are talking about a fourth *spatial* dimension. A fuller description of our universe would require the addition of a time dimension onto whatever spatial dimensions one is considering.

Try to come up with ways to help the 2-D person imagine what happens to the mitten when it is taken out of the plane into 3-space. It may help to think of intersecting planes rotating with respect to each other: How will a 2-D person in one of the planes experience it? Draw on the person's experience living in two dimensions, as well as some of your own experiences and attempts to imagine four dimensions.

*PROBLEM **20.2**   A 3-SPHERE IN 4-SPACE

We will now explore 3-dimensional spheres in 4-space, which is possibly the shape of our physical universe. We include the following terminology to help clarify the terms and parameters of later problems in this chapter.

> **DEFINITIONS:** Let $\mathbf{R}^4$ be the collection of 4-tuples of real numbers (x,y,z,w) with the distance function (metric)
>
> $$d(\,(a,b,c,d),\,(e,f,g,h)\,) = \sqrt{(a-e)^2 + (b-f)^2 + (c-g)^2 + (d-h)^2}\,.$$
>
> $\mathbf{R}^4$ can be considered as a 4-dimensional vector space or, with the metric, as 4-dimensional Euclidean space.

Note that every plane in $\mathbf{R}^3$ has exactly one line perpendicular to it at every point. A line is ***perpendicular*** to a plane if it intersects the plane and is perpendicular to every line in the plane that passes through the intersection point. In $\mathbf{R}^4$ we can similarly

a. *Show that every 2-dimensional subspace (a plane containing the origin O), Π, in $\mathbf{R}^4$ has an **orthogonal complement**, $\Pi^\perp$, which is a 2-dimensional subspace (plane) that intersects Π only at O such that every line through O in Π is perpendicular to every line through O in $\Pi^\perp$.*

This was probably proved in your linear algebra course. The easiest proof (I think) is to change the orthonormal basis in $\mathbf{R}^4$ so that, in the new coordinates, Π is the span of the first two basis vectors.

> **DEFINITIONS:** Let a 3-sphere, $\mathbf{S}^3$, be the collection of points in $\mathbf{R}^4$ that are at a fixed distance r from O, the center of $\mathbf{R}^4$. The number r is called the **radius** of the sphere.
>
> We define a **great circle** on $\mathbf{S}^3$ to be the intersection of $\mathbf{S}^3$ with a plane in $\mathbf{R}^4$ through the center O.
>
> We define a **great 2-sphere** on $\mathbf{S}^3$ to be the intersection of $\mathbf{S}^3$ with any 3-dimensional subspace of $\mathbf{R}^4$ (that passes through O.)

  b. *Show that every great 2-sphere in the 3-sphere has reflection-in-itself symmetry.*

Choose an orthonormal basis for $\mathbf{R}^4$ so that the great 2-sphere is in the 3-subspace spanned by the first three basis elements.

  c. *Show that every great circle has the symmetries in $\mathbf{S}^3$ of rotation through any angle and reflection through any great 2-sphere perpendicular to the great circle. Because these are principle symmetries of a straight line in 3-space, it makes sense to call these great circles **geodesics in $\mathbf{S}^3$**.*

Choose an orthonormal basis for $\mathbf{R}^4$ so that the great circle is in the plane spanned by the first two basis elements.

  d. *If two great circles in $\mathbf{S}^3$ intersect, then they lie in the same great 2-sphere.*

SUGGESTIONS FOR PROBLEM 20.2

Thinking in four dimensions may be a foreign concept to you, but believe it or not, it is possible to visualize a 4-dimensional space. Remember, the fourth dimension here is not time, but a fourth spatial dimension. We know that any two intersecting lines that are linearly independent (that do not coincide) determine a 2-dimensional plane. If we then add another line that is not in this plane, the three lines span a 3-space. When lines such as these are used as coordinate axes for a coordinate system, then they are typically taken to be orthogonal — each line is perpendicular to the others. Now to get 4-space, imagine a fourth line that is perpendicular to each of these original three. This creates the fourth dimension that we are considering.

Although we cannot experience all four dimensions at once, we can easily imagine any three at a time, and we can easily draw a picture of any two. This is the secret to looking at four dimensions. These 3- or 2-dimensional subspaces look exactly the same as any other 3-space or plane that you have seen before. This holds true for any subspace of 4-space — because all four of the coordinate lines are orthogonal, any set of three of these will look the same and will determine a space geometrically identical to our familiar 3-space, and any set of two coordinate lines will look like any other and will determine a 2-dimensional plane.

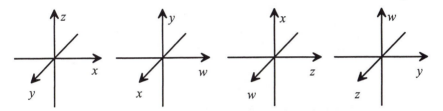

Figure 20.5   Any three coordinate axes determine a 3-space

For all of these problems, you should not be looking at projections of the 3-sphere into a plane or a 3-space, but rather looking at the part of the 3-sphere that lies in a subspace. For example, because the 3-sphere is defined as the set of points a distance r from the origin O in $\mathbf{R}^4$, if you take any 3-dimensional subspace (through O) of $\mathbf{R}^4$, then the part of the 3-sphere that lies in this 3-dimensional subspace is the set of points a distance r from its center O in the 3-space. So any 3-dimensional subspace of $\mathbf{R}^4$ intersects the 3-sphere in a 2-sphere, which you know all about by now, and you can easily visualize.

For all of the problems here, it is generally best to draw pictures of various planes (2-dimensional subspaces) through the 3-sphere because they are easy to draw on a piece of paper. Remember, only include in your picture those geometric objects that lie in the plane you are drawing. So, a great circle that lies in this plane would be drawn as a circle, while another great circle that passed through this plane would intersect this plane only in two points. See Figure 20.6.

For this particular problem, you are looking at the 3-sphere extrinsically. A good way to proceed is to draw several planes as outlined above, and try to get an idea of how the planes relate to one another when combined into a 4-dimensional space. Once you have an understanding of how the different planes interact in four dimensions, it is fairly easy to show how the great circles of a 3-sphere behave.

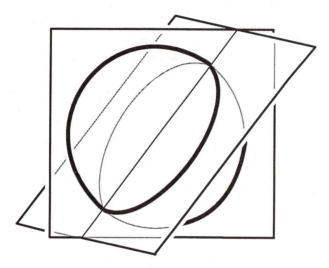

Figure 20.6   Intersecting great circles

*PROBLEM 20.3   HYPERBOLIC 3-SPACE, UPPER HALF SPACE

As previously mentioned, there is no smooth isometric embedding of a hyperbolic plane in 3-space and, thus, no analytic isometric description. In the same way there is no isometric analytic description of a hyperbolic 3-space in 4-space. Instead we will describe hyperbolic 3-space intrinsically in terms of the upper half space model that is analogous to the upper half plane model.

DEFINITION: Let $\mathbf{R}^{3+} = \{(x,y,z)$ in $\mathbf{R}^3 \mid z > 0\}$ and call it *the upper half space*.

In Chapter 5 we started with the annular hyperbolic plane and then defined a coordinate map $z: \mathbf{R}^{2+} \rightarrow H^2$. Now we do not have an isometric model $\mathbf{H}^3$ but, instead, we have to start with the upper half space and use z to define $\mathbf{H}^3$. Recall that z: $\mathbf{R}^{2+} \to \mathbf{H}^2$ has distortion ρ/b at the point (a,b) in $\mathbf{R}^{2+}$, where ρ is the radius of the annuli. As we saw in Chapter 5 we can study the geometry of $\mathbf{H}^2$ by considering it to be the upper half plane with angles as they are in $\mathbf{R}^{2+}$ and distances distorted in $\mathbf{R}^{2+}$ by ρ/b at the point (a,b). So now we use this idea to define $\mathbf{H}^3$.

> **DEFINITION:** Define *the upper half space model* of hyperbolic space $\mathbf{H}^3$ to be the upper half space $\mathbf{R}^{3+}$ with angles as they are in $\mathbf{R}^{3+}$ and with distances distorted by ρ/c at the point (a,b,c). We call ρ the *radius* of H^3.
>
> We define a *great semicircle* in $\mathbf{H}^3$ to be the intersection of $\mathbf{H}^3$ with any circle that is in a plane perpendicular to the boundary of $\mathbf{R}^{3+}$ and whose center is in the boundary of $\mathbf{R}^{3+}$ or the intersection of $\mathbf{R}^{3+}$ with any line perpendicular to the boundary of $\mathbf{R}^{3+}$. The *boundary of* $\mathbf{R}^{3+}$ are those points in $\mathbf{R}^3$ with $z = 0$.
>
> We define a *great hemisphere* in $\mathbf{R}^{3+}$ to be the intersection of $\mathbf{R}^{3+}$ with a sphere whose center is on the boundary of $\mathbf{R}^{3+}$ in $\mathbf{R}^3$ or the intersection of $\mathbf{R}^{3+}$ with any plane that is perpendicular to the boundary of $\mathbf{R}^{3+}$ in $\mathbf{R}^3$.

a. *Show that inversion through a great hemisphere in $\mathbf{R}^{3+}$ has distortion 1 in $\mathbf{H}^3$ and, thus, is an isometry in $\mathbf{H}^3$ and can be called a (hyperbolic) reflection through the great hemisphere.*

Look back at Problem **5.3**. Note that any inversion in a sphere when restricted to a plane containing the center of the sphere is an inversion of the plane in the circle formed by the intersection of the plane and the sphere.

b. *Show that, given a great semicircle [or great hemisphere], there is a hyperbolic reflection (inversion through a great hemisphere) that takes the great semicircle [hemisphere] to a vertical half line [half plane] in the upper half space.*

Look back at Problems **14.2d** and **16.3**. Note that any inversion in a sphere when restricted to a plane containing the center of the sphere is an inversion of the plane in the circle formed by the intersection of the plane and the sphere.

Any vertical half plane is precisely an Upper Half Plane Model of $\mathbf{H}^2$. Thus we conclude that each great hemisphere in $\mathbf{H}^3$ has the geometry of $\mathbf{H}^2$.

*c. *Show that every great semicircle has the symmetries in $\mathbf{H}^3$ of reflection through any great hemisphere perpendicular to the great semicircle and rotation about the great semicircle through any angle. Because these are principle symmetries of a straight line in 3-space it makes sense to call these great semicircles **geodesics in** $\mathbf{H}^3$.*

For the reflection, use the arguments of Problem **14.2b**. For the rotation, refer back to Problem **6.2b** and restrict your attention to the vertical half plane that passes through the center of Γ and is perpendicular to Γ.

d. *If two great semicircles in $\mathbf{H}^3$ intersect, then they lie in the same great hemisphere.*

Use Part **b** to assume that one of the great semicircles is a vertical half line.

*PROBLEM 20.4 DISJOINT EQUIDISTANT GREAT CIRCLES

a. *Show that there are two great circles in $\mathbf{S}^3$ such that **every** point on one is a distance of one-fourth of a great circle away from **every** point on the other and vice versa.*

Is there anything analogous to this in $\mathbf{H}^3$ or in ordinary 3-space? Why?

SUGGESTIONS

This problem is especially interesting because there is no equivalent theorem on the 2-sphere; we know that on the 2-sphere, all great circles intersect, so they can't be everywhere equidistant. The closest analogy on the 2-sphere is that a pole is everywhere equidistant from the equator. When we go up to the next dimension, this pole "expands" to a great circle such that every point on this great circle is everywhere equidistant from the equator. While this may seem mind-boggling, there are ways of seeing what is happening.

The main difference created by adding the fourth dimension lies in the orthogonal complement to a plane. In 3-space, the orthogonal complement of a plane is a line that passes through a given point. This means that for any given point on the plane, (the origin is always a convenient point), there is exactly one line that is perpendicular to the plane at that point. Now, what happens when you add the fourth dimension? In 4-space, the orthogonal complement to a plane is a plane. This means that every line in one plane is perpendicular to every line in the other plane. To understand how this is possible, think about how it works in 3-space and refer back to Figure 20.5. Now look at the xy-plane and the zw-plane. What do you notice? Why is every line through the center in one of these planes perpendicular to every line through the center in the other?

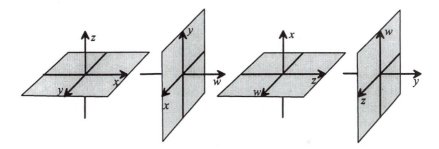

Figure 20.7   Orthogonal planes

Knowing this, look at the two great circles in terms of the planes in which they lie, and look at the relationships between these two planes, that is, where and how they intersect. Also, try to understand how great circles can be everywhere equidistant.

If we rotate along a great circle on a 2-sphere, all points of the sphere will move except for the two opposite poles of the great circle. If you rotate along a great circle on a 3-sphere, then the whole 3-sphere will move except for those points that are a quarter great circle away from the rotating great circle. Therefore, if you rotate along one of the two great circles you found above, the other great circle will be left fixed. But now rotate the 3-sphere simultaneously along both great circles at the same speed. Now every point is moved and is moved along a great circle!

b. *Write an equation for this rotation and check that each point of the 3-sphere is moved at the same speed along some great circle. Show that all of the great circles obtained by this rotation are equidistant from each other* (in the sense that the perpendicular distance from every point on one great circle to another of the great circles is a constant).

These great circles are traditionally called ***Clifford parallels***. See [**DG:** Thurston], pp 103–04, and [**DG:** Penrose] for readable discussions of Clifford parallels. These great circles are named after William Clifford (1845–1879, English).

*PROBLEM **20.5** HYPERBOLIC AND SPHERICAL SYMMETRIES

We are now ready to see that the symmetries of great circles and great 2-spheres in a 3-sphere [and great semicircles and great hemispheres in a hyperbolic 3-space] are the same as the symmetries of straight lines and (flat) planes in 3-space. If g is a great circle in the 3-sphere, then let $g^{\perp}$ denote the great circle (from Problem **20.4**) every point of which is $\pi/2$ from every point of g.

a. *Check the entries in the table* (Figure 20.8), *which gives a summary of various symmetries of lines, great circles, and great semicircles and of (flat) planes, great 2-spheres, and great hemispheres.*

DEFINITION: A surface in a 3-sphere or in a hyperbolic 3-space is called ***totally geodesic*** if, for any every pair of points on the surface, there is a geodesic (with respect to $\mathbf{S}^3$ or $\mathbf{H}^3$) that joins the two points and also lies entirely in the surface.

b. *Show that a great 2-sphere in $\mathbf{S}^3$ (with radius r) is a totally geodesic surface and is itself a sphere of the same radius r.*

c. *Show that a great hemisphere is a totally geodesic surface in $\mathbf{H}^3$ (with radius r) and is isometric to a hyperbolic plane with the same radius r.*

In the upper half space model there is a hyperbolic reflection that takes every great hemisphere to a plane perpendicular to the boundary. (See Problem **20.3**.)

symmetries of...	reflection through...	reflection through...	half-turn about...	rotation about...	translation along...
line $l \subset R^2$	l	line $\perp l$	point in l	NA	l
great circle $g \subset S^2$	g	gr. circle $\perp g$	pt/pair in g	poles of g	g
gr. s-circle $g \subset H^2$	g	gr. s-circle $\perp g$	pt/pair in g	NA	g
line $l \subset R^3$	plane $\supset l$	plane $\perp l$	line $\perp l$ intersecting l	l	l
great circle $g \subset S^3$	gr. sphere $\supset g$	gr. sphere $\perp g$	gr. circle $\perp g$ intersecting g	g	g
gr. s-circle $g \subset H^3$	gr. h-sphere $\supset g$	gr. h-sphere $\perp g$	gr. s-cir. $\perp g$ intersecting g	g	g
plane $P \subset R^3$	P	plane $\perp P$	line in P	line $\perp P$	line $\subset P$
great sphere $G \subset S^3$	G	gr. sphere $\perp G$	gr. circle in G	gr. circle $\perp G$	great circle $\subset G$
gr. h-sphere $G \subset S^3$	G	gr. h-sphere $\perp G$	gr. s-circle in G	gr. s-circle $\perp G$	gr. s-circle $\subset G$

Figure 20.8   Symmetries in Euclidean, spherical, and hyperbolic spaces

PROBLEM **20.6**   TRIANGLES IN **3**-DIMENSIONAL SPACES

Show that if A, B, C are three points in S^3 *[or in* H^3*] that do not all lie on the same geodesic, then there is a unique great 2-sphere [hemisphere],* G^2*, containing A, B, C.*

Thus, we can define △ABC as the (small) triangle in G² with vertices A, B, C. With this definition, **triangles in S³ [or in H³] have all the properties that we have been studying of small triangles on a sphere [or triangles in a hyperbolic plane].**

SUGGESTIONS

Think back to the suggestions in Problems **20.2** and **20.3** — they will help you here, as well. Take two of the points, A and B, and show that they lie on a unique plane through the center, O, of the 3-sphere [or a unique plane perpendicular to the boundary of $\mathbf{R}^{3+}$]. Then show that there is a unique (shortest) geodesic in this plane.

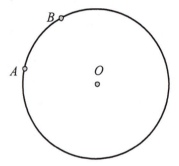

Figure 20.9   Great circle through A and B

Think of A, B, and C as defining three intersecting great circles [or semicircles]. On a 3-sphere, look at the planes in which these great circles lie, and where the two planes lie in relation to one another. In hyperbolic 3-space, use a hyperbolic reflection to send one of the great semicircles to a vertical line.

Be sure to show that the great 2-sphere (hemisphere) containing A, B, C is unique.

Chapter 21

POLYHEDRA

> ... if four equilateral triangles are put together, three of
> their plane angles meet to form a single solid angle,... When
> four such angles have been formed the result is the simplest
> solid figure...
>
> The second figure is composed of ... eight equilateral
> triangles, which yield a single solid angle from four planes.
> The formation of six such solid angles completes the second
> figure.
>
> The third figure ... has twelve solid angles, each
> bounded by five equilateral triangles, and twenty faces, each of
> which is an equilateral triangle.
>
> ... Six squares fitted together complete eight solid
> angles, each composed by three plane right angles. The figure
> of the resulting body is the cube ...
>
> There is still remained a fifth construction, which the
> god used for arranging the constellations on the whole heaven.
>
> — Plato, *Timaeus*, 54e–55c [**AT: Plato**]

DEFINITIONS AND TERMINOLOGY

A *tetrahedron*, $\triangle ABCD$, in 3-space [in a 3-sphere or a hyperbolic 3-space][†] is determined by any four points, A, B, C, D, called its *vertices*, such that all four points do not lie on the same plane [great 2-sphere, great hemisphere] and no three of the points lie on the same line [geodesic]. The *faces* of the tetrahedron are the four [small] triangles $\triangle ABC$, $\triangle BCD$, $\triangle CDA$, $\triangle DAB$. The *edges* of the tetrahedron are the six line

---

[†]The text in the brackets applies to polyhedra on a 3-sphere or a hyperbolic 3-space.

[geodesic] segments *AB, AC, AD, BC, BD, CD*. The **interior** of the tetra-hedron is the [smallest] 3-dimensional region that it bounds.

Tetrahedra are to 3-dimensions as triangles are to 2-dimensions. Every polyhedron can be dissected into tetrahedra, but the proofs are considerably more difficult than the ones from Problem **7.5b**, and in the discussion to Problem **7.5** there is a polyhedron that is impossible to dissect into tetrahedra without adding extra vertices. There are numerous congruence theorems for tetrahedra, analogous to the congruence theo-rems for triangles. All of the problems below apply to tetrahedra in Euclidean 3-space or a 3-sphere or a hyperbolic 3-space.

The **dihedral angle**, ∠*AB*, at the edge *AB* is the angle formed at *AB* by Δ*ABC* and Δ*ABD*. The dihedral angle is measured by intersecting it with a plane that is perpendicular to *AB* at a point between *A* and *B*. The **solid angle** at *A*, ∠*A*, is that portion of the interior of the tetrahedron "at" the vertex *A*. See Figure 21.1.

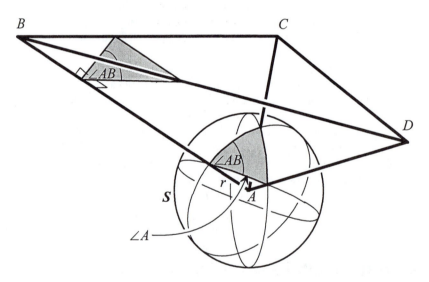

Figure 21.1   Dihedral and solid angles

PROBLEM 21.1   MEASURE OF A SOLID ANGLE

The measure of the solid angle is defined as the ratio

$$\mathrm{m}(\angle A) = [\lim_{R \to 0}] \text{ area}\{ \text{ (interior of } \Delta ABCD) \cap S \} / r^2 ,$$

where S is any small 2-sphere with center at A whose radius, r, is smaller than the distance from A to each of the other vertices and to each of the edges and faces not containing A.

    a. *Show that the measures of the solid and dihedral angles of a tetrahedron satisfy the following relationship:*

$$m(\angle A) = m(\angle AB) + m(\angle AC) + m(\angle AD) - \pi.$$

    b. *Show that two solid angles with the same measure are not necessarily congruent.*

SUGGESTIONS

Solid angles, whether in Euclidean 3-space or a 3-sphere or a hyperbolic 3-space, are closely related to spherical triangles on a small sphere around the vertex. You can think of starting with a sphere, S, and creating a solid angle by extending three sticks out from the center of the sphere. If you connect the ends of these sticks, you will have a tetrahedron. The important thing to notice is how the sticks intersect the sphere. They will obviously intersect the sphere at three points, and you can draw in the great circle arcs connecting these points. Look at the planes in which the great circles lie. In this problem you need to figure out the relationships between the angles of the spherical triangle and the dihedral angles.

    The formula given above for the definition of the solid angle uses the intersection of the interior of the solid angle with any small sphere S. This intersection is the small triangle that you just drew, and the area of the intersection is the area of the triangle. Because the measure of a solid angle is defined in terms of an area, it is possible for two solid angles to have the same measure without being congruent — they can have the same area without having the same shape.

    What you are asked to prove here is the relationship between the measure of a solid angle and the measures of its dihedral angles. Because they are closely related to spherical triangles on the small sphere, you can use everything you know about small triangles on a sphere.

PROBLEM **21.2** EDGES AND FACE ANGLES

We will study congruence theorems for tetrahedra that can be thought of as the three-dimensional analogue of triangles. A tetrahedron has 4 vertices, 4 faces, and 6 edges and we can denote it by $\triangle ABCD$ where A, B, C, D are the vertices.

> *If $\triangle ABCD$ and $\triangle A'B'C'D'$ are two tetrahedra such that*
>
> $\angle BAC \cong \angle B'A'C'$, $\angle CAD \cong \angle C'A'D'$, $\angle BAD \cong \angle B'A'D'$,
> $CA \cong C'A'$, $BA \cong B'A'$, $DA \cong D'A'$
>
> *then*
>
> $\triangle ABCD \cong \triangle A'B'C'D'$.

Part of your proof must be to show that the solid angles $\angle A$ and $\angle A'$ are congruent and not merely that they have the same measure.

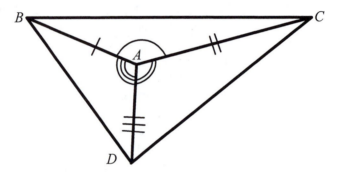

Figure 21.2   Edges and faces

SUGGESTIONS

If S is a small sphere with center at A and radius r, then

$$S \cap (\text{interior of } \triangle ABCD)$$

is a spherical triangle whose sides have lengths

$$r\angle BAC, \ r\angle CAD, \ r\angle BAD.$$

In the last problem, you saw how solid angles are related to spherical triangles. This problem asks you to prove the congruence of tetrahedra based on certain angle and length measurements. (Note that the angles shown above are not the dihedral angles of the tetrahedron.) So, since you can use spherical triangles to relate solid and dihedral angle measurements, why not use them to prove tetrahedra congruencies? Use the hint given to see what measurements of the spherical triangle are defined by measurements of the tetrahedron. Then see if the measurements given do in fact show congruence, and show why.

PROBLEM **21.3** EDGES AND DIHEDRAL ANGLES

If

$$AB \cong A'B', \ \angle AB \cong \angle A'B', \ AC \cong A'C',$$
$$\angle AC \cong \angle A'C', \ AD \cong A'D', \ \angle AD \cong \angle A'D',$$

then

$$\triangle ABCD \cong \triangle A'B'C'D'.$$

This is very similar to the previous problem, but uses different measurements — here we have the dihedral angles instead of the angles on the faces of tetrahedron. Look at this problem the same way you looked at the previous one — see how the measurements given relate to a spherical triangle, and then prove the congruence.

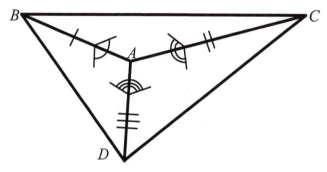

Figure 21.3   Edges and dihedral angles

PROBLEM **21.4** OTHER TETRAHEDRA CONGRUENCE THEOREMS

*Make up your own congruence theorems! Find and prove at least two other sets of conditions that will imply congruence for tetrahedra; that is, make up and prove other theorems like those in Problems **21.2** and **21.3**.*

It is important to make sure your conditions are sufficient to prove that the solid angles are congruent, not just that they have the same measure.

PROBLEM 21.5 THE FIVE REGULAR POLYHEDRA

A *regular polygon* is a polygon lying in a plane or 2-sphere or hyperbolic plane such that all of its edges are congruent and all of its angles are congruent. For example, on the plane a regular quadrilateral is a square. On a 2-sphere and a hyperbolic plane a regular quadrilateral is constructed as shown in Figure 21.4. See also Figure 17.21 for a regular octagon on a hyperbolic plane.

Note that half of a regular quadrilateral is a Khayyam quadrilateral (see Chapter 12). On 2-spheres and hyperbolic planes there are no similar polygons; for example, a regular quadrilateral (congruent sides and congruent angles) will have the same angles as another regular quadrilateral if and only if they have the same area. (Do you see why?)

A *polyhedron* in 3-space [or in a 3-sphere or in a hyperbolic 3-space] is *regular* if all of its edges are congruent, all of its face angles are congruent, all of its dihedral angles are congruent, and all of its solid angles are congruent. The faces of a polyhedron are assumed to be polygons that lie on a plane [a great 2-sphere, a great hemisphere].

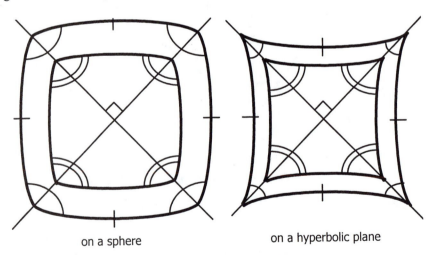

on a sphere   on a hyperbolic plane

Figure 21.4   Regular quadrilaterals

Show that there are only five regular polyhedra. In Euclidean 3-space, to say "there are only five regular polyhedra" is to mean that any regular polyhedra is similar (same shape, but not necessarily the same size) to one of the five. It still makes sense on a 3-sphere and a hyperbolic 3-space to say that "there are only five regular polyhedra," but you need to make clear what you mean by this phrase.

These polyhedra are often called "the Platonic Solids," and are described by Greek philosopher Plato (429–348 B.C.) as "forms of bodies which excel in beauty" (*Timaeus*, 53e [**AT:** Plato]), but there is considerable evidence that they were known well before Plato's time. See T. L. Heath's discussion in [**AT:** Euclid, *Elements*], Vol. 3, pp. 438–39, for evidence that the five regular solids were known by Greeks before the time of Plato. In addition, there is described in [**Hi:** Critchlow], pp. 148–49, the discovery in Scotland of a complete set of the five regular polyhedra carefully carved out of stone by Neolithic persons some 4,000 to 6,000 years ago. The regular polyhedra are also the subject of the thirteenth (and last) book in [**AT:** Euclid, *Elements*].

SUGGESTIONS

Your argument should be essentially the same whether you are considering 3-space, or a 3-sphere, or a hyperbolic 3-space. There are many widely different ways to do this problem. Following we suggest some approaches:

First Approach: Note that the faces of a regular polyhedron must be regular polygons. Then focus on the vertices of regular polyhedra. Show that if the faces are regular quadrilaterals or regular pentagons, then there must be precisely three faces intersecting at each vertex. Show that it is impossible for regular hexagons to intersect at a vertex to form the solid angle of a regular polyhedron. If the faces are regular (equilateral) triangles, then show that there are three possibilities at the vertices.

Second Approach: Refer to Problem 17.6. Each regular polyhedron can be considered to be projected out from its center onto a sphere and thus determine a cell-division of the sphere. The Euler number of this spherical subdivision is $v - e + f = 2$, where v is the number of vertices, e is the number of edges, and f is the number of faces. Then,

$$2e = nf \text{ and } 2e = kv . \ (\text{Why?})$$

Thus, deduce that

$$e = \frac{2}{\frac{2}{k} + \frac{2}{n} - 1},$$

and remember that e must be a positive integer.

In both approaches you should then finish the problem by using earlier problems from this chapter to show that any two polyhedra constructed from the same polygons, with the same number intersecting at each vertex, must be congruent. This step is necessary because there are polyhedra that are not rigid (that is, there are polyhedra that can be continuously moved into a non-congruent polyhedra without changing any of the faces or changing the number of faces coming together at each vertex. See Robert Connelly's "The Rigidity of Polyhedral Surfaces" (*Mathematics Magazine*, vol. 52, no. 5 (1979), pp. 275–83).

The five regular polyhedra are usually named the Tetrahedron, the Cube, the Octahedron, the Dodecahedron, and the Icosahedron. (See Figure 21.5.) There is a duality (related to but not exactly the same as the duality in Chapter 19, *Trigonometry and Duality*) among regular polyhedra: If you pick the centers of the faces of a regular polyhedron, then these points are the vertices of a regular polyhedron, which is called the **dual** of the original polyhedron. You can see that the cube is dual to the octahedron (and vice versa), that the icosahedron is dual to the dodecahedron (and vice versa), and that the tetrahedron is dual to itself.

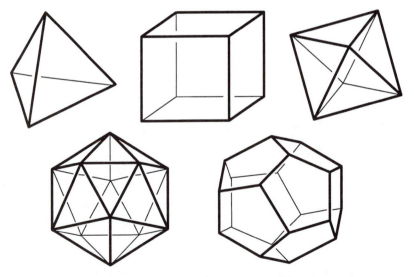

Figure 21.5   The five Platonic solids

Chapter 22

3-MANIFOLDS — THE SHAPE OF SPACE

> ...if we look at the extreme points of the sky, all the visual rays appear equal to us, and if diametrically opposed stars describe a great circle, one is setting while the other is rising. If the universe, instead of being spherical, were a cone or a cylinder, or a pyramid or any other solid, it would not produce this effect on earth: one of its parts would appear larger, another smaller, and the distances from earth to heaven would appear unequal.
> — Theon of Smyrna (~70–~135, Greek), [**AT:** Theon]

> It will be shown that a multiply extended quantity [three-dimensional manifold] is susceptible of various metric relations, so that Space constitutes only a special case of a triply extended quantity. From this however it is a necessary consequence that the theorems of geometry cannot be deduced from general notions of quantity, but that those properties that distinguish Space from other conceivable triply extended quantities can only be inferred from experience.
> — G.F.B. Riemann (1826–1866, German) *On the Hypotheses Which Lie at the Foundations of Geometry*, translated in [**DG:** Spivak], Vol. II, p.135.

Now we come to where we live. We live in a physical 3-dimensional space, that is (at least) locally like Euclidean 3-space. The fundamental question we will investigate in this chapter is: How can we tell what the shape of our Universe is? This is the same question both Theon of Smyrna and Riemann were attempting to answer in the above quotes. This is a

287

very difficult question for which there is currently (as this is written) no clear answer. However, there are several things we can say.

SPACE AS AN ORIENTED GEOMETRIC 3-MANIFOLD

Presumably our physical Universe is globally a geometric 3-manifold. (A *geometric 3-manifold* is a space in which each point in the Universe has a neighborhood that is isometric with a neighborhood of either Euclidean 3-space, a 3-sphere, or a hyperbolic 3-space. The notion of a 3-manifold was introduced in the work by Riemann quoted above.) I say "is globally a geometric 3-manifold" in the same sense in which we say that globally Earth is a sphere (and spherical geometry is the appropriate geometry for intercontinental airplane flights) even though it is clear almost anywhere on the earth that locally there are many hills and valleys that make Earth not locally isometric to a sphere. However, the highest point on Earth (Mount Everest) is 8.85 km above sea level and the lowest point on the floor of the ocean (the Mariana Trench) is 10.99 km below sea level — the difference is only about 0.3% of the 6368 km radius of Earth (variations in the radius are of the same magnitude).

It is known that locally our physical Universe is definitely NOT a geometric 3-manifold. It was predicted by Einstein's general theory of relativity that the local curvature of our physical Universe is affected by any mass (especially large masses like our sun and other stars) [Albert Einstein (1879–1955), German/Swiss/American]. This effect is fairly accurately illustrated by imagining a 2-dimensional universe that is the surface of a flat rubber sheet. If you place steel balls on this rubber sheet the balls will locally make dents or dimples in the sheet and thus will locally distort the flat Euclidean geometry. Einstein's prediction has been confirmed in two ways, as follows:

1. The orbits of the planets Mercury, Jupiter, and Saturn are (quite accurately) ellipses, and the major axes of these elliptical orbits change directions (precess). Classical Newtonian mechanics (based on Euclidean geometry) predicts that in a century the precession will be

    Mercury: 1.48°, Jupiter: 1.20°, Saturn: 0.77°,

    measured in degrees of an arc.

Astronomers noticed that the observed amount of precession agreed accurately with these values for Saturn and Jupiter; however, for Mercury (the closest planet to the Sun) the observed precession is 1.60°, which is 0.12° more than is predicted by Newtonian/Euclidean methods. But, if one does the computations based on the curvature of space near the sun that is predicted by Einstein, the calculations agree accurately with the observed precession. For a mathematician's description of this calculation (which uses formulas from differential geometry), see Frank Morgan's *Riemannian Geometry: A Beginner's Guide* [**DG:** Morgan], Chapter 7.

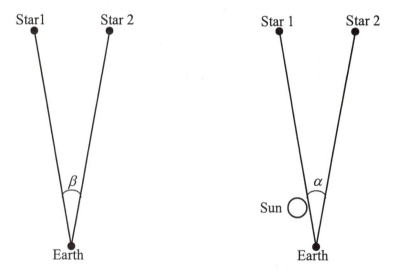

Figure 22.1   Observing local non-Euclidean geometry in the Universe

2. In 1919 British astronomers led by Arthur Eddington measured the angle subtended by two stars from Earth: once when the Sun was not near the path of the light from the stars to Earth and once when the path of light from one of the stars went very close to the Sun. See Figure 22.1. In order to be able to see the star when its light passes close to the Sun they had to make the second observation during a total eclipse of the Sun. They observed that the angle (α) measured with the Sun near was smaller than the angle (β) measured when the Sun was not near. The difference between the two measured angles was exactly

what Einstein's theory predicted. Some accounts of this experiment talk about the Sun "bending" the light rays, but it is more accurate to say that light follows intrinsically straight paths (geodesics) and that the Sun distorts the geometry (curvature) of the space nearby. For more details of this experiment and other experiments that verified Einstein's theory, see

http://www.ncsa.uiuc.edu/Cyberia/NumRel/EinsteinTest.html

But, most of our physical Universe is empty space with only scattered planets, stars, galaxies, and the "dimples" that the stars make in space is only a very small local effect near the star. The vast empty space appears to locally (on a medium scale much larger than the scale of the distortions near stars) be a geometry of the local symmetries of Euclidean 3-space. (In particular, our physical space is observed to be locally (on a medium scale) the same in all directions with isometric rotations, reflections, and translations as in Euclidean 3-space.

THEOREM 22.0. *Euclidean 3-space, 3-spheres, and hyperbolic 3-spaces are the only simply connected* (every loop can be continuously shrunk to a point in the space) *3-dimensional geometries that locally have the same symmetries as Euclidean 3-space.*

The condition of simply connected is to rule out general geometric manifolds modeled on Euclidean 3-space, 3-spheres, or hyperbolic 3-spaces. There is a discussion of the proof (and more precise statement) of this theorem in [**DG:** Thurston], Section 3.8, where Thurston discusses eight possible 3-dimensional simply connected geometries, but only three have the same symmetries as Euclidean 3-space. A more elementary discussion without proofs can be found in [**DG:** Weeks], Chapter 18.

Unfortunately (or maybe fortunately!), geometric 3-manifolds are not fully understood. At this point no one knows what all of the geometric 3-manifolds are or how to distinguish one from the other. The theory of 3-manifolds is an area of current active research. For a very accessible discussion of this research see [**DG:** Weeks], Part III. For detailed discussions of this research see Thurston's *Three-Dimensional Geometry and Topology*, Volume 1, [**DG:** Thurston] and the second (and further?) volumes as soon as they appear.

PROBLEM **22.1** IS OUR UNIVERSE NON-EUCLIDEAN?

There is a widely reported story about the famous mathematician Carl Friedrich Gauss (1777–1855, German). According to the story he tried to measure the angles of a triangle whose vertices were three mountain peaks in Germany. If the sum of the angles had turned out to be other than 180°, then he would have deduced that the Universe is not Euclidean (or that light does not travel in straight lines). However, his measurements were inconclusive because he measured the angles at 180° within the accuracy of his measuring instruments. This story is apparently a myth [see Breitenberger, "Gauss' Geodesy and the Axiom of Parallels," *Archive for the History of the Exact Sciences*, 31(1984), pp. 273–89], for though he did measure the angles of a large triangle of mountain peaks, the purpose was to connect several grids of triangles that were being used for making a complete survey of Europe. Nevertheless, the story still leads to the question

a. *Could we now show that the Universe is non-Euclidean by measuring the angles of a large triangle in our solar system? How accurately would we have to measure the angles?*

Note that, if the Universe is a 3-sphere or a hyperbolic 3-space, the radius R of the Universe would have to be at least as large as the diameter of our galaxy, which is about 10^{18} km. In the foreseeable future, the largest triangle whose angles we could measure has area less than the area of our solar system, which is about 8×10^{19} km^2. Use the formulas you found in Problems **7.1** and **7.2** and extended in Problem **12.6**.

So, in order to determine the geometry of space we have to look further than our solar system.

b. *If the stars were distributed uniformly in space* (it is not clear to what extent this is actually true), *how could you tell by looking at stars at different distances whether space was locally Euclidean, spherical, or hyperbolic?*

If you have trouble envisioning this then start with the analogous problem for a 2-dimensional bug on a plane, sphere, or hyperbolic plane. What would this bug observe? Assume that you can tell how far away each star is — this is something that astronomers know how to do.

  c. *Suppose you know that certain types of stars (or galaxies) have a fixed known amount of brightness (astronomers call such stars or galaxies, **standard candles**), and you can see several of these standard candles at various distances from Earth. How could you tell whether the Universe is Euclidean, spherical, or hyperbolic?*

The apparent brightness of a shining object in Euclidean space is inversely proportional to the square of the distance to the object.

  d. *It is impractical to measure the excess of triangles in our solar system by only taking measurements of angles within our solar system. What observations of only distant stars and galaxies would tell us that the Universe is not Euclidean 3-space?*

If you have trouble conceptualizing a 3-sphere or hyperbolic 3-space, then you can do this problem, first, for a very small bug on a 2-sphere or hyperbolic plane who can see distant points (stars), but who is restricted to staying inside its "solar system," which is so small that any triangle in it has excess (or defect) too small to measure. Always think intrinsically! You can assume, generally, that light will travel along geodesics, so think about looking at various objects and the relationships you would expect to find. For example, if the Universe were a 3-sphere and you could see all the way around the Universe (the distance of a great circle), how would you know that the Universe is spherical? Why? What if we could see half way around the Universe? Or a quarter of the way around? Think of looking at stars at these distances.

  All the ways discussed in Parts **a–d** have been tried by astronomers, but, up to now, none of these observations has been accurate enough to make a definite determination.

  Recently, astronomers have attempted to find at a far (but approximately known) distance in the Universe a structure whose size is known (based on physical assumptions). If such a structure is found then it is possible to measure the angle subtended by this structure from Earth. If the geometry of the Universe is Euclidean then the measure of the observed angle is predicted by the Law of Cosines. If the observed angle is larger than the prediction, then the Universe would be spherical and if it is smaller, then the Universe would be hyperbolic. (*Do you see why?*)

In spring 2000, a group of astronomers announced (see editorial and article in *Nature*, volume 404, April 2000) that they have observed such a structure in the cosmic microwave background radiation and that their observations are "consistent with a flat, euclidean Universe." But, even if later observations and analysis determine that the Universe is Euclidean, it will still leave open the question of which Euclidean 3-manifold the Universe is. We will study Euclidean 3-manifolds in the next section.

There is a satellite scheduled for a year 2000 launch (and another more accurate one scheduled for a launch in 2007) that is designed to make an accurate map of the cosmic microwave background radiation. An analysis of this map may provide the clues we need to definitely determine the global geometry of space. To describe how this will work we must first investigate geometric 3-manifolds.

PROBLEM **22.2** EUCLIDEAN **3-MANIFOLDS**

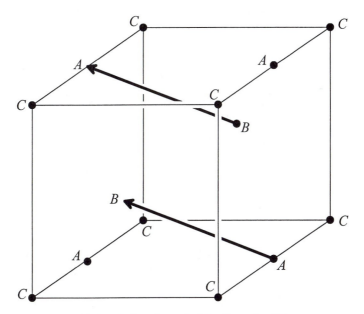

Figure 22.2 A closed geodesic path on the 3-torus

We now consider the 3-dimensional analogue of the flat torus. Consider a cube in Euclidean 3-space with the opposite faces glued through a reflection in the plane that is midway between the opposite faces. See

Figure 22.2. In this figure, I have drawn a closed straight path which starts from *A* on the bottom right edge and then hits the middle of the front face at *B*. It continues from the middle of the back face and finishes at the middle of the top left edge at a point that is glued to *A*.

a. *Show that the cube with opposite faces glued by a reflection through the plane midway between is a Euclidean 3-manifold. That is, check that a neighborhood of each point is isometric to a neighborhood in Euclidean 3-space. This Euclidean 3-mani-fold is call the* **3-*torus*.**

Look separately at points (such as *A*) that are in the middle of edges, points (such as *B*) that are the middle of faces, and points that are verti-ces (such as *C*).

   Next glue the vertical faces of the cube the same way but glue the top and bottom faces by a quarter turn. In this case we get the closed geodesic depicted in Figure 22.3.

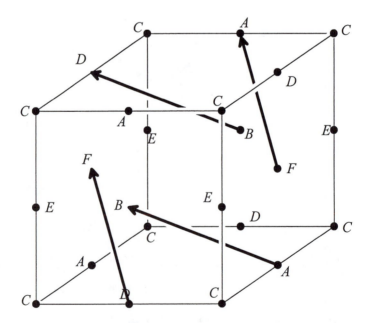

Figure 22.3   A closed geodesic in the quarter turn manifold

b. *Show that you obtain a Euclidean 3-manifold from the cube with vertically opposite faces glued by a reflection through the plane midway between, and the top and bottom faces glued by a quarter turn rotation.* This Euclidean 3-manifold is called the **quarter turn manifold**.

c. *Draw a picture similar to Figures 22.2 and 22.3, for the **half turn manifold**, which is the same as the quarter turn manifold except that it is obtained by gluing the top and bottom faces with a half turn.*

In Problem **17.2** we represented the flat torus in two different ways — one starting with a rectangle or square and the other starting from a hexagon. The above discussion of the 3-torus corresponds to the construction of the flat torus from a square. Now we want to look at what happens if we use an analogue of the hexagon construction.

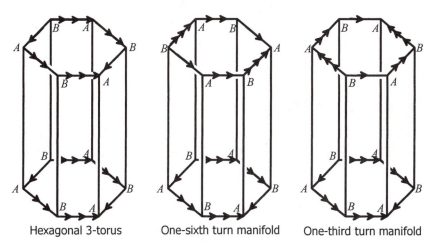

| Hexagonal 3-torus | One-sixth turn manifold | One-third turn manifold |

Figure 22.4   Hexagonal 3-manifolds

Consider a hexagonal prism as in Figure 22.4. We will make gluings on the vertical sides by gluing each vertical face with its opposite in such a way that each horizontal cross-section (which are all hexagons) has the same gluings as the hexagonal flat torus (see Problem **17.2b** and Figure 17.8). The top and bottom face we glue in one of three ways. If we glue the top and bottom face through a reflection in the halfway plane then we obtain the **hexagonal 3-torus**. If we glue the top and bottom faces with a one-sixth rotation, then we obtain the ***one-sixth turn manifold***. If we glue the top and bottom with a one-third rotation then we will get the ***one-third turn manifold***.

> **d.** *Show that the hexagonal 3-torus, the one-sixth turn manifold, and the one-third turn manifold are Euclidean 3-manifolds and that the hexagonal 3-torus is homeomorphic to the 3-torus. What happens if we consider the two-thirds turn manifold and the three-sixths turn manifold and the five-sixth turn manifold?*

It can be shown that

> **THEOREM 22.2.** *There are exactly ten Euclidean 3-manifolds up to homeomorphism. Of these four are non-orientable and six are orientable. Five of the six orientable Euclidean manifolds are the 3-torus, the quarter turn manifold, the half turn manifold, the one-sixth turn manifold, and the one-third turn manifold.*

See [**DG:** Weeks], page 252, for a discussion of this theorem. For more detail and a proof see [**DG:** Thurston], Section 4.3.

PROBLEM 22.3 DODECAHEDRAL 3-MANIFOLDS

Spherical and hyperbolic 3-manifolds are more complicated than Euclidean 3-manifolds. In fact, no one knows what all the hyperbolic 3-manifolds are. We will only look at a few examples in order to get an idea of how to construct spherical and hyperbolic 3-manifolds in this problem and the next.

There are two examples can be obtained by making gluings of the faces of a dodecahedron (see Problem **21.5**). It will be best for this problem for you to have a model of the dodecahedron that you can look at and touch. We want to glue the opposite faces of the dodecahedron. Looking at your dodecahedron (or Figure 22.5) you should see that the opposite faces are not lined up but rather are rotated one-tenth of a full turn from each other. Thus there are three possibilities for gluings: We can glue with a one-tenth rotation, or a three-tenths rotation, or a five-tenths (= one-half) rotation. When making the rotations it is important to always rotate in the same direction (say clockwise as you are facing the face from the outside). *You should check with your model that this is the same as rotating clockwise while facing the opposite face.*

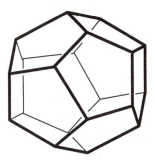

Figure 22.5   Dodecahedron

a. *When you glue the opposite faces of a dodecahedron with a one-tenth clockwise rotation, how many edges are glued together? What if you use a three-tenths rotation? Or a one-half rotation?*

I find the best way to do this counting is to take my model and mark one edge with tape. Then for each of the two pentagon faces on which the edge lies, the gluing glues the edge to another edge on the other side of the dodecahedron — mark those edges also. Continue with those marked edges until you have marked all the edges that are glued together.

Your answers to Part **a** should be 2, 3, 5 (NOT in that order!). Thus in order for these manifolds to be geometric manifolds we must use dodecahedrons with dihedral angles of 180°, 120°, and 72° in either Euclidean space, a sphere, or a hyperbolic 3-space. But before we go further we must figure out the size of the dihedral angles of the dodecahedron in Euclidean space, which from our model seems to be close to (if not equal to) 120°.

b. *Calculate the size ϕ of the dihedral angle of the (regular) dodecahedron in Euclidean 3-space.*

Imagine a small sphere with center at one of the vertices of the dodecahedron. This sphere will intersect the dodecahedron in a spherical equilateral triangle. This triangle is called the ***link of the vertex*** in the dodecahedron. Determine the lengths of the sides of this triangle and then use the Law of Cosines (Problem **19.2**).

Now imagine a very small dodecahedron in 3-sphere. Its dihedral angles will be very close to the Euclidean angle ϕ (*Why is this the case?*). If you now imagine the dodecahedron growing in the 3-sphere, its dihedral angles will grow from ϕ. If you start with a very small dodecahedron in a hyperbolic space then its dihedral angles will start very close to ϕ and then decrease as the dodecahedron grows.

    c. *Show that the manifold from Part **a** with three edges being glued together is a spherical 3-manifold if $\phi < 120°$, or a Euclidean 3-manifold if $\phi = 120°$, or a hyperbolic 3-manifold if $\phi > 120°$. This geometric 3-manifold is called the **Poincaré dodecahedral space** in honor of Henri Poincaré (1854–1912, French) who first described (not using the dodecahedron) a space homeomorphic to this geometric 3-manifold.*

Show that each vertex of the dodecahedron is glued to three other vertices and that the four solid angles fit together to form a complete solid angle in the model (either Euclidean 3-space, 3-sphere, or hyperbolic 3-space).

    d. *Can the dihedral angles of a dodecahedron in a 3-sphere grow enough to be 180°? What does such a dodecahedron look like? Is the manifold with two edges being glued together a spherical 3-manifold? This spherical 3-manifold is called the **projective 3-space** or **RP³**.*

    e. *Can the dihedral angles of a dodecahedron in a hyperbolic 3-space shrink enough to be equal to 72°? If so, the dodecahedral manifold with five edges being glued together is a hyperbolic 3-manifold. This hyperbolic 3-manifold is called the **Seifert-Weber dodecahedral space**, after H. Seifert and C. Weber, who first described both dodecahedral spaces in a 1933 article.[†]*

Imagine that the dodecahedron grows until its vertices are at infinity (thus on the bounding plane in the upper half space model). Use the fact

---

[†] "Die beiden Dodekaederräume," *Mathematische Zeitschrift*, v. 37 (1933), no. 2, p. 237.

that angles are preserved in the upper half space model and look at the three great hemispheres that are determined by the three faces coming together at a vertex. Remember to also check that the solid angles at the vertices of the dodecahedron fit together to form a complete solid angle.

PROBLEM **22.4** SOME OTHER GEOMETRIC **3-MANIFOLDS**

We now look at three more examples of geometric 3-manifolds.

a. *Start with a tetrahedron and glue the faces as indicated in Figure 22.6. Does this gluing produce a manifold? Can the tetrahedron be put in a 3-sphere or hyperbolic 3-space so that the gluings produce a geometric 3-manifold? We call this the* ***tetrahedral space.***

Investigate how many edges are glued together and what happens near the vertices.

Figure 22.6   Tetrahedral space

b. *Start with a cube and glue each face to the opposite face with a one-quarter turn rotation. Does this gluing produce a manifold? Can the cube be put in a 3-sphere or hyperbolic 3-space so that the gluings produce a geometric 3-manifold? This is called the* ***quaternionic manifold*** *because its symmetries can be expressed in the quaternions (a four-dimensional version of the complex numbers with three imaginary axes and one real axis).*

Again, investigate how many edges are glued together and what happens near the vertices.

> **c.** *Start with an octahedron and glue each face to the opposite face with a one-sixth turn rotation. Does this gluing produce a manifold? Can the octahedron be put in a 3-sphere or hyperbolic 3-space so that the gluings produce a geometric 3-manifold?* This is called the **octahedral space**.

Again, investigate how many edges are glued together and what happens near the vertices. You can also use your knowledge of solid angles from Chapter 21.

Cosmic Background Radiation

Astronomers from earthbound observatories have noticed a radiation that is remarkably uniform coming to Earth from all directions of space. In 1991, the USA's Cosmic Background Explorer (COBE) mapped large portions of this radiation to a resolution of about 10 degrees of arc. COBE determined that the radiation is uniform to nearly one part in 100,000, but there are slight variations (or texture) observed. It is this texture that gives us the possibility to determine the global shape of the Universe. To understand how this determination may be possible we must first understand from where (and when!) this background radiation came.

The generally accepted explanation of the cosmic background radiation is that in the early stages of the Universe matter was so dense that no radiation could escape (the space was filled with matter so dense that all radiation was scattered). But at a certain point in time (about 300,000 years after the big bang and about 13 billion years ago) matter in the Universe started coalescing and the density of matter decreased enough that radiation could start traveling though the Universe in all directions, but was still dense enough that the radiating matter (radiating because it was hot) was fairly homogeneously distributed throughout the Universe. It is this first escaped radiation that we see when we look at the cosmic background radiation.

Remember that all radiation (light and others) travels at the speed of light ($\sim 10^{13}$ kilometers per year) and thus the cosmic radiation that reaches us today has been traveling for about 13 billion years and has traveled about 1.3×10^{22} km. Thus the cosmic background radiation gives us a picture of a sphere that was a slice of the early Universe roughly 13 billion years ago. The cosmologists call this the ***last scattering surface***. It appears to us that Earth is at the center of this sphere, but this is talking in space-time. The physical center of this sphere is the point in the early Universe where 13 billion years later the Milky Way galaxy, solar system, and Earth would form.

The cosmologists say that the above discussion does not involve any assumptions about the density of matter or the presence of a "cosmological constant," both of which are hotly debated subjects among cosmologists. In addition, the only assumption made about the geometry of our physical Universe is the assumption that our physical Universe is a geometric manifold and that, even though it is expanding, the topology is constant (or at least has been constant since the Universe was big enough for the radiation to escape). In Theorem **22.5**, we see that in fact if we know the topology we also know the geometry.

Before going on with the 3-dimensional discussion, let us look at an analogous situation in 2-dimensions. Imagine that the 2-dimensional bug's universe is a flat torus obtained from a square with opposite side glued. We may assume that the bug is at the center of the square looking out in all directions at a textured circle from the center of that circle, the last scattering circle for the bug. If this circle has diameter larger than the side of the square then the circle will intersect itself as indicated in Figure 22.7.

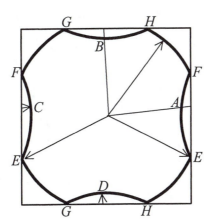

Figure 22.7   Seeing the 2-dimensional scattering circle

Note, in Figure 22.7, that when the bug in the center looks toward the point A the bug will not see A but rather will see C. (The light from A will have reached the center of the square earlier!) Likewise, the bug will see the point D on the circle but not the point B. The important points to focus on are the points, *E*, *F*, *G*, *H*, where the circle intersects itself. The bug will see these points in two different directions. See Figure 22.7 where the point *E* is seen from both sides. If the texture is unique enough, the bug should be able to tell that it is looking at the same point in two different directions. The pattern of these identical point-pairs will indicate to the bug that its Universe is the flat torus. (See Problem **22.5a**.)

The three-dimensional situation is similar. Consider that our physical Universe is a 3-torus that is the result of gluings on a cube as in Figure 22.2. If Earth is considered to be in the center of this cube and if the sphere of last scatter has reached the faces of the cube then the intersection of the sphere of last scatter with the faces of the cube will be circles and (because of the gluings) the circle on one face will be identified (by a reflection) with the circle on the opposite face (see Figure 22.8). The pattern of circles shows the underlying cube and the gluings. In Problem **22.5b** you will explore what the pattern of circles will look like in the other geometric 3-manifolds we have discussed.

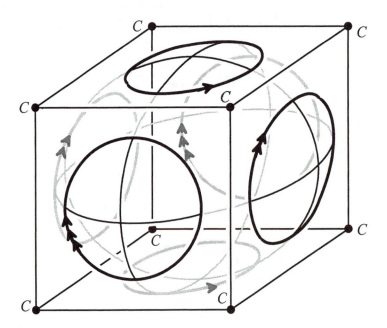

Figure 22.8   Self-intersections of the sphere of last scatter in the 3-torus

In our physical Universe, there are two space probes that are scheduled to be launched to study the texture of the microwave background radiation. In Fall 2000 (about the time that this book is published) the USA's National Aeronautic and Space Administration plans to launch the Microwave Anisotropy Probe (MAP). See

http://map.gsfc.nasa.gov/

for more information. In 2007, the European Space Agency is scheduled to launch the Planck satellite. See

http://astro.estec.esa.nl/SA-general/Projects/Planck/planck.html

for more information. These probes will produce detailed maps of the texture of the microwave radiation. MAP should produce 0.3° resolution and the Planck satellite is hoped to provide about 0.1° resolution. Remember that the current maps have only 10° resolution.

For further discussion of these measurements see Luminet, Starkman, and Weeks, "Is Space Finite?", *Scientific American*, April 1999, page 90; or Cornish and Weeks, "Measuring the Shape of the Universe," *Notices of the American Mathematical Society*, 1998, or look for more recent articles.

PROBLEM 22.5 CIRCLE PATTERNS SHOW THE SHAPE OF SPACE

*a. *For each of the geometric 2-manifolds in Problems 17.1, 17.2, 17.4, 17.5, what would the patterns of matching point-pairs look like if the bug's last scattering circle was large enough to intersect itself?*

Draw pictures analogous to Figure 22.7.

b. *For each of the geometric 3-manifolds in Problems 22.2, 22.3, and 22.4, what would the pattern of matching circles look like if our physical Universe were the shape of that manifold **and** if the sphere of first scatter reaches far enough around the Universe for it to intersect itself.*

Draw and examine, as best you can, pictures analogous to Figure 22.8.

We saw in Problem **17.6** that the area of a spherical or hyperbolic 2-manifold is determined by its topology and the radius (or curvature) of the model. Similarly, (but much more complicatedly) we have for spherical and hyperbolic 3-manifolds,

> **THEOREM 22.5.** *If two orientable spherical or hyperbolic 3-manifolds are homeomorphic then they are geometrically similar. That is, two such homeomorphic manifolds are isometric if the model spherical or hyperbolic spaces have the same radius (curvature).*

Thus, if we are successful in finding a pattern of circles that determines the Universe is a spherical or hyperbolic 3-manifold, then we will know the volume of our physical Universe.

Watch the news media and science and mathematics journals for news as analysis of the data from Microwave Anisotropy Probe and Planck satellite start coming in (probably in 2001). As they come in and I learn about them, I will post updates of these probes and the analyses of the data that are relevant to this text at the website

http://www.math.cornell.edu/~dwh/books/eg00

Appendix A

Euclid's Definitions, Postulates, and Common Notions

At the age of eleven, I began Euclid, with my brother as my tutor. This was one of the great events of my life, as dazzling as first love. I had not imagined that there was anything so delicious in the world.

— Bertrand Russell (1883),
Autobiography: 1872–1914, Allen & Unwin, 1967, p. 36

The following are the definitions, postulates, common notions listed by Euclid in the beginning of his *Elements*, Book 1. These are Heath's translations from [**AT:** Euclid, *Elements*] except that I modified them to make the wording and usage more in line with word usage today. In my modifications I used Heath's extensive notes on the translation in order not to change the meanings involved. But, remember that we can not be sure of the exact meaning intended by Euclid — any translation should be considered only as an approximation.

Definitions

1. *A **point** is that which has no parts.*

2. *A **curve** is length without width.*
   [Heath translates this as "line," but today we normally use the term "*curve*" in place of "line".]

3. *The ends of a curve are points.*
   [It is not assumed here that the curve *has* ends (for example, see Definition 15 below); but, if the line does have ends then the ends are points.]

4. *A (straight) line is a curve that lies symmetrically with the points on itself.*
   [The commonly quoted Heath's translation says "...lies *evenly* with the points...", but in his notes he says "we can safely say that the sort of idea which Euclid wished to express was that of a line ... without any irregular or unsymmetrical feature distinguishing one part or side of it from another."]

5. *A surface is that which has length and width only.*

6. *The boundaries of a surface are curves.*

7. *A plane is a surface that lies symmetrically with the straight lines on itself.*
   [The comments for Definition 4 apply here as well.]

8. *An angle is the inclination to one another of two curves in a plane that meet on another and do not lie in a straight line.*
   [What Euclid meant by the term "inclination" is not clear to me and apparently also to Heath.]

9. *The angle is called rectilinear when the two curves are straight lines.*
   [Of course, we (and Euclid in most of the *Elements*) call these simply "angles".]

10. *When a straight line intersects another straight line such that the adjacent angles are equal to one another, then the equal angles are called right angles and the lines are called perpendicular straight lines.*
    [As discussed in the last section of Chapter 4, cones give us examples of spaces where right angles (as defined here) are not always equal to 90 degrees.]

11. *An obtuse angle is an angle greater than a right angle.*

12. *An acute angle is an angle less than a right angle.*

13. *A **boundary** of anything is that which contains it.*

14. *A **figure** is that which is contained by any boundary or boundaries.*

15. *A **circle** is a plane figure contained by one curve (called the **circumference**) with a given point lying within the figure such that all the straight lines joining the given point to the circumference are equal to one another.*

16. *The given point of a circle is called the **center** of the circle.*

17. *A **diameter** of a circle is any straight line drawn through the center and with its end points on the circumference, and the straight line bisects the circle.*

18. *A **semicircle** is a figure contained by a diameter and the part of the circumference cut off by it. The **center** of the semicircle is the same as the center of the circle.*

19. ***Polygons** are those figures whose boundaries are made of straight lines: **triangles** being those contained by three, **quadrilaterals** those contained by four, and **multilaterals** those contained by more than four straight lines.*

20. *An **equilateral triangle** is a triangle that has three equal sides, an **isosceles triangle** is a triangle that has only two of its sides equal, and a **scalene triangle** is a triangle that has all three sides unequal.*

21. *A **right triangle** is a triangle that has a right angle, an **obtuse triangle** is a triangle that has an obtuse angle, and an **acute triangle** is a triangle that has all of its angles acute.*

22. *A **square** is a quadrilateral that is equilateral (has all equal sides) and right angled (has all right angles), a **rectangle** is a quadrilateral that is right angled but not equilateral, a **rhombus** is a quadrilateral that is equilateral but not right angled, a **rhomboid** is a quadrilateral that has opposite sides and angles equal to one another but that is neither equilateral nor right angled. Let quadrilaterals other than these be called **trapezia**.*

23. **Parallel** *straight lines are straight lines lying in a plane, which do not meet if continued indefinitely in both directions.*

POSTULATES

1. *A (unique) straight line may be drawn from any point to any other point.*

2. *Every limited straight line can be extended indefinitely to a (unique) straight line.*

3. *A circle may be drawn with any center and any distance.*

4. *All right angles are equal.*
   [Note that cones give us examples of spaces in which all right angles are not equal, see Chapter 4. Thus this postulate could be rephrased: *"There are no cone points."*]

5. *If a straight line intersecting two straight lines makes the interior angles on the same side less than two right angles, then the two lines (if extended indefinitely) will meet on that side on which the angles are less than two right angles.*
   [See Chapter 10 for more discussion of this postulate.]

COMMON NOTIONS

1. *Things that are equal to the same thing are also equal to one another.*

2. *If equals are added to equals, then the results are equal.*

3. *If equals are subtracted from equals, the remainders are equal.*

4. *Things that coincide with one another are equal to one another.*

5. *The whole is greater than any of its parts.*

Appendix B

SQUARE ROOTS IN THE SULBASUTRAM[†]

Like the crest of a peacock, like the gem on the head of a snake, so is mathematics at the head of all knowledge.
— Vendanga Jyotisa (about 500 B.C.)

INTRODUCTION

In Chapter 13 we described the "divide and average" (D&A) method for approximating the numerical value of a square root. This method is also sometimes called "the Archytas/Newton's method" and seems to be the most widely used and most efficient modern numerical method. On a January 1990 visit to the Sankaracharya Mutt in Konchipuram, Tamilnadu, India, I was surprised to discover that there is a method based on Baudhayana's *Sulbasutram* [**AT:** Baudhayana] which produces for many square roots the same approximations as the D&A method but with significantly fewer computations. "Sulbasutram" means "rules of the cord" and the several Sanskrit texts collectively called the *Sulbasutra* were written by the Vedic Hindus starting before 600 B.C. and are thought to be compilations of oral wisdom that may go back to 2000 B.C. (See, for example, A. Seidenberg *The Ritual Origin of Geometry* [**Hi:** Seidenberg].) These texts have prescriptions for building fire altars, or *Agni*. However, contained in the *Sulbasutra* are sections that constitute a geometry textbook detailing the geometry necessary for designing and

---

[†]Much of the material in this appendix will also appear in a chapter by the author in *Geometry at Work*, MAA Notes, volume 53 [**SE:** Gorini].

constructing the altars. As far as I have been able to determine these are the oldest geometry (or even mathematics) textbooks in existence. There are at least four versions of the *Sulbasutram* by Baudhayana, Apastamba, Katyayana, and Manava. The geometric descriptions are very similar in these four books and I will only use Baudhayana's version here.

Baudhayana's Sutra 52 states that the diagonal of a square is the side of another square with two times the area of the first square. Thus, if we consider the side of the original square to be one unit, then the diagonal is the side (or root) of a square of area two, or simply the square root of 2, that is $\sqrt{2}$. The Sanskrit word for this length is *dvi-karani* or, literally, "that which produces 2".

Sutras 61 and 62 contain the following prescription for finding the length of the diagonal of a square:

> Increase the length [of the side] by its third and this third by its own fourth less the thirty-fourth part of that fourth. The increased length is a small amount in excess (*saviśeṣa*).[†]

Thus, the above passage from the *Sulbasutram* gives the approximation

$$\sqrt{2} \approx 1 + \tfrac{1}{3} + \tfrac{1}{4} \cdot \tfrac{1}{3} - \tfrac{1}{34} \cdot \tfrac{1}{4} \cdot \tfrac{1}{3}.$$

I use $\approx$ instead of $=$ indicating that the Vedic Hindus were aware that the length they prescribed is a little too long (*saviśeṣa*). In fact my calculator gives

$$\sqrt{2} \approx 1.4142135 \cdots$$

and the *Sulbasutram*'s value expressed in decimals is

$$\sqrt{2} \approx 1.4142156 \cdots$$

So the question arises — how did the Vedic Hindus obtain such an accurate numerical value? Unfortunately, there is nothing that survives which records how they arrived at this *saviśeṣa*.

There have been several speculations as to how this value was obtained. See [**GC**: Datta] for a discussion of several of these, some of which are also discussed in [**Hi**: Joseph]. But no one as far as I can determine has noticed that there is a step-by-step method (based on

---

[†]This last sentence is translated by some authors as "The increased length is called *saviśeṣa*". I follow the translation of "*saviśeṣa*" given in [**GC**: Datta], pp. 196–202; see also [**Hi**: Joseph] who translates the word as "a special quantity in excess".

geometric techniques in the *Sulbasutram*) that will not only obtain the approximation

$$\sqrt{2} \approx 1 + \tfrac{1}{3} + \tfrac{1}{4} \cdot \tfrac{1}{3} - \tfrac{1}{34} \cdot \tfrac{1}{4} \cdot \tfrac{1}{3},$$

but can also be continued indefinitely to obtain as accurate an approximation as one wishes.

This method will in one more step obtain

$$\sqrt{2} \approx 1 + \tfrac{1}{3} + \tfrac{1}{4} \cdot \tfrac{1}{3} - \tfrac{1}{34} \cdot \tfrac{1}{4} \cdot \tfrac{1}{3} - \tfrac{1}{1154} \cdot \tfrac{1}{34} \cdot \tfrac{1}{4} \cdot \tfrac{1}{3},$$

where the only numerical computation needed is $1154 = 2[(34)(17) - 1]$ and, moreover, the method shows that the square of this approximation is less than 2 by exactly

$$\frac{1}{(1154 \cdot 34 \cdot 4 \cdot 3)^2} = \frac{1}{221{,}682{,}772{,}224}.$$

The interested reader can check that this approximation is accurate to eleven decimal places.

In the *Sulbasutram* the *agni* are described as being constructed of bricks of various sizes. Mention is made in Sutra 3–7 of square bricks of side 1 *pradesa* (span of a hand, about 9 inches) on a side. Each *pradesa* was equal to 12 *angula* (finger width, about 3/4 inch) and one *angula* was equal to 34 sesame seeds laid together with their broadest faces touching. Thus the diagonal of a *pradesa* brick had length

1 *pradesa* + 4 *angula* + 1 *angula* − 1 sesame thickness.

I do not believe it is purely by chance that these units come out this nicely. Notice that this length is too large by roughly one-thousandth of the thickness of a sesame seed. Presumably there was no need for more accuracy in the building of altars!

Baudhayana's *Sulbasutram* does not show how he found the *saviśeṣa*. But let us see if his geometric methods discussed above hint at a method for finding numerical approximations of square roots.

CONSTRUCTION OF THE *SAVIŚEṢA* FOR THE SQUARE ROOT OF 2

If we apply Sutra 54 to the union of two squares each with sides of 1 *pradesa* we get a square with side 1½ *pradesa* from which a square of side ½ *pradesa* had been removed. See the Figure 14.6.

Now we can attempt to take a strip from the left and bottom of the large square — the strips are to be just thin enough that they will fill in the little removed square. The pieces filling in the little square will have length 1/2, and six of these lengths will fit along the bottom and left of the large square. The reader can then see that strips of thickness

$$(1/6)(1/2) \; pradesa \; (= 1 \; angula)$$

will (almost) work.

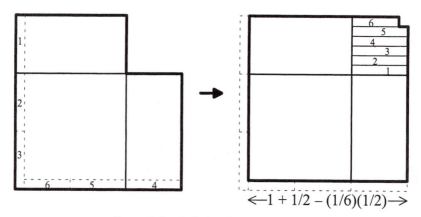

$$\longleftarrow 1 + 1/2 - (1/6)(1/2) \longrightarrow$$

Figure B.1   Refining the construction

There is still a little square left out of the upper right corner because the thin strips overlapped in the lower left corner.

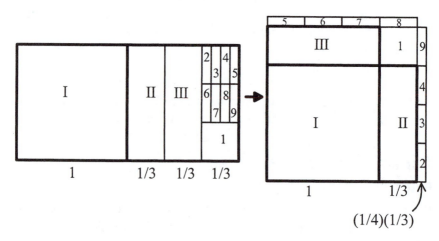

$$(1/4)(1/3)$$

Figure B.2   Alternative construction of first approximation

Notice that

$$\left(1 + \tfrac{1}{2} - \tfrac{1}{6} \cdot \tfrac{1}{2}\right)\text{pradesa} = 1\ \text{pradesa} + 5\ \text{angulas}$$

$$= \left(1 + \tfrac{1}{3} + \tfrac{1}{4} \cdot \tfrac{1}{3}\right)\text{pradesa}$$

We can get directly to $1 + \tfrac{1}{3} + \tfrac{1}{4} \cdot \tfrac{1}{3}$ by considering the dissection in Figure B.2.

We now have that two square *pradesas* are equal to a large square minus a small square. The large square has side equal to 1 *pradesa* plus 1/3 of a *pradesa* plus 1/4 of 1/3 of a *pradesa*, or 1 *pradesa* and 5 *angulas*, and the small square has side of 1 *angula*. To make this into a single square we may attempt to remove a thin strip from the left side and the bottom just thin enough that the strips will fill in the little square. Because these two thin strips will have length 1 *pradesa* and 5 *angulas* or 17 *angulas*, we may cut each into 17 rectangular pieces each 1 *angula* long. If these are stacked up they will fill the little square if the thickness of the strips is 1/34 of an *angula* (or $\tfrac{1}{34} \cdot \tfrac{1}{4} \cdot \tfrac{1}{3}$ *pradesa*). Without a microscope we will now see the two square *pradesas* as being equal in area to the square with side

$$1 + \tfrac{1}{3} + \tfrac{1}{4} \cdot \tfrac{1}{3} - \tfrac{1}{34} \cdot \tfrac{1}{4} \cdot \tfrac{1}{3}\ \text{pradesa}.$$

But with a microscope we see that the strips overlap in the lower left corner and, thus, that there is a tiny square of side $\tfrac{1}{34} \cdot \tfrac{1}{4} \cdot \tfrac{1}{3}$ still left out. See Figure B.3.

Thus, $1 + \tfrac{1}{3} + \tfrac{1}{4} \cdot \tfrac{1}{3} - \tfrac{1}{34} \cdot \tfrac{1}{4} \cdot \tfrac{1}{3}$ is still a little in excess. We can now perform the same procedure again by removing a very very thin strip from the left and bottom edges and then cutting them into $\tfrac{1}{34} \cdot \tfrac{1}{4} \cdot \tfrac{1}{3}$ *pradesa* lengths in order to fill in the left out square. If w is twice the number of $\tfrac{1}{34} \cdot \tfrac{1}{4} \cdot \tfrac{1}{3}$ lengths in $1 + \tfrac{1}{3} + \tfrac{1}{4} \cdot \tfrac{1}{3} - \tfrac{1}{34} \cdot \tfrac{1}{4} \cdot \tfrac{1}{3}$ *pradesa*, then the strips we remove must have width $\tfrac{1}{2k} \cdot (\tfrac{1}{34} \cdot \tfrac{1}{4} \cdot \tfrac{1}{3})$ *pradesa*. We can calculate w easily because we already noted that there were 17 segments of length $\tfrac{1}{4} \cdot \tfrac{1}{3}$ in the length $1 + \tfrac{1}{3} + \tfrac{1}{4} \cdot \tfrac{1}{3}$ and each of these segments was divided into 34 pieces and then one of these pieces was removed. Thus, $w = 2[34(17) - 1] = 1154$ and

$$\sqrt{2} \approx 1 + \tfrac{1}{3} + \tfrac{1}{4} \cdot \tfrac{1}{3} - \tfrac{1}{34} \cdot \tfrac{1}{4} \cdot \tfrac{1}{3} - \tfrac{1}{1154} \cdot \tfrac{1}{34} \cdot \tfrac{1}{4} \cdot \tfrac{1}{3}$$

with error expressed by $2 \cdot 1 =$

$$(1 + \tfrac{1}{3} + \tfrac{1}{4} \cdot \tfrac{1}{3} - \tfrac{1}{34} \cdot \tfrac{1}{4} \cdot \tfrac{1}{3} - \tfrac{1}{1154} \cdot \tfrac{1}{34} \cdot \tfrac{1}{4} \cdot \tfrac{1}{3})^2 - (\tfrac{1}{1154} \cdot \tfrac{1}{34} \cdot \tfrac{1}{4} \cdot \tfrac{1}{3})^2.$$

I write "2·1" instead of "2" to remind us that for Baudhayana (and, in fact, for most mathematicians up until near the end of the 19th century) $\sqrt{2}$ denoted the side (a *length*) of a square with *area* 2.

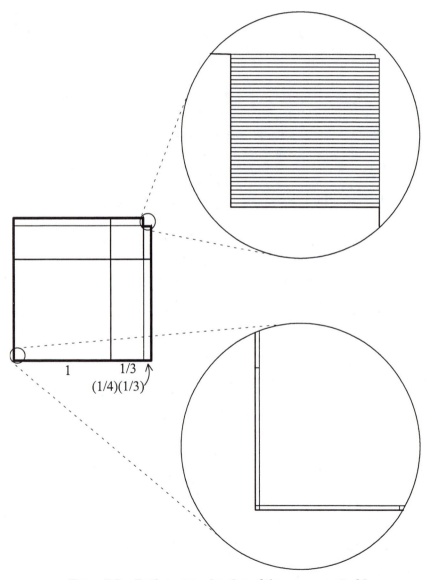

Figure B.3   Further approximation of the square root of 2

If we again follow the same procedure of removing a very thin strip from the left and bottom edges and cutting them into $\frac{1}{1154} \cdot \frac{1}{34} \cdot \frac{1}{4} \cdot \frac{1}{3}$ length pieces, then the reader can check that the number of such pieces must be

$$2[1154(1154/2) - 1] = (1154)^2 - 2 = 1{,}331{,}714$$

and, thus, that the next approximation (*saviśeṣa*) is

$$1 + \tfrac{1}{3} + \tfrac{1}{4} \cdot \tfrac{1}{3} - \tfrac{1}{34} \cdot \tfrac{1}{4} \cdot \tfrac{1}{3} - \tfrac{1}{1154} \cdot \tfrac{1}{34} \cdot \tfrac{1}{4} \cdot \tfrac{1}{3} - \tfrac{1}{1{,}331{,}714} \cdot \tfrac{1}{1154} \cdot \tfrac{1}{34} \cdot \tfrac{1}{4} \cdot \tfrac{1}{3}$$

The difference between 2·1 and the square of this *saviśeṣa* is

$$\left(\tfrac{1}{1{,}331{,}714} \cdot \tfrac{1}{1154} \cdot \tfrac{1}{34} \cdot \tfrac{1}{4} \cdot \tfrac{1}{3} \right)^2.$$

This method will work for any number N that you can first express as the area of the difference of two squares, $N \cdot 1 = A^2 - B^2$, where the side A is an integral multiple of the side B. For example,

$$5 \cdot 1 = (2 + \tfrac{1}{4})^2 - (\tfrac{1}{4})^2, \ 7 \cdot 1 = (2 + \tfrac{2}{3})^2 - (\tfrac{1}{3})^2, \ 10 \cdot 1 = (3 + \tfrac{1}{6})^2 - (\tfrac{1}{6})^2,$$

$$12 \cdot 1 = (3 + \tfrac{1}{2})^2 - (\tfrac{1}{2})^2, \ \text{and} \ (2 + \tfrac{1}{2}) \cdot 1 = (1 + \tfrac{1}{2} + \tfrac{1}{6} \cdot \tfrac{1}{2})^2 - (\tfrac{1}{6} \cdot \tfrac{1}{2})^2.$$

I find that the easiest way for me to see that these expressions are valid is to represent them geometrically in a way that would also have been natural for Baudhayana. To illustrate these geometric representations see Figure B.4.

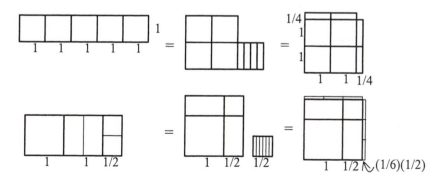

Figure B.4   Finding *saviśeṣas* for 5 and 2½

Figures 14.13 and 14.14 give other examples. The reader should try out this method to see how easy it is to find *saviśeṣas* for the square roots of other numbers, for example, 3, 11, 2¾.

FRACTIONS IN THE SULBASUTRAM

You have probably noticed that all the fractions above are expressed as unit fractions, but this is not always the case in the Baudhayana's *Sulbasutram*. For example, in Sutra 69 he discusses how to find a length that is an approximation to the diagonal of a square whose side is the "third part of" 8 *prakramas* (which equals 240 *angulas*). He describes the construction:

> ... increase the measure [the 8 *prakramas*] by its fifth, divide the whole into five parts and make a mark at the end of two parts.

In more modern notation if we let D equal 8 *prakramas*, then this gives the approximation of the diagonal of a square with side $(1/3)D$ as

$$\tfrac{2}{5}(D + \tfrac{1}{5}D).$$

This is equivalent to $\sqrt{2}$ being approximated by 1.44.

If you attempt to find the *saviśeṣas* for other square roots you will find it convenient to use non-unit fractions. For example, start with the picture in Figure B.5.

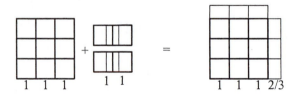

Figure B.5   Finding the square root of 13

You can make slight modifications in the above method to find

$$13 \cdot 1 = (3 + \tfrac{2}{3} - \tfrac{1}{11} \cdot \tfrac{2}{3} - \tfrac{1}{120} \cdot \tfrac{1}{11} \cdot \tfrac{2}{3})^2 - (\tfrac{1}{120} \cdot \tfrac{1}{11} \cdot \tfrac{2}{3})^2.$$

COMPARING WITH THE
DIVIDE-AND-AVERAGE (D&A) METHOD

Today the most efficient method usually taught to find square roots is called "divide-and-average." If you wish to find the square root of N then you start with an initial approximation a_0 and then take as the next approximation the average of a_0 and N/a_0. In general, if a_n is the n-th approximation of the square root of N, then $a_{n+1} = \frac{1}{2}(a_n + (N/a_n))$. The interested reader can check that if you start with $[1+(1/3)+(1/12)] = [17/12] = 1.416666666667$ as your first approximation of $\sqrt{N}$, then the succeeding approximations are numerically the same as those given by Baudhayana's geometric method.

D&A — Calculator	D&A — Fractions
$a_1 = 1.416666667$	$a_1 = 17/12$
$a_2 = \frac{1}{2}(a_1 + (2/a_1)) =$ 1.414215686	$a_2 = \frac{1}{2}[(17/12)+ 2(12/17)] =$ (577/408)
$a_3 = \frac{1}{2}(a_2 + (2/a_2)) =$ 1.414213562	$a_3 = \frac{1}{2}[(577/408)+2(408/577)] =$ (665857/470832)
$a_4 = \frac{1}{2}(a_3 + (2/a_3)) =$ 1.414213562	$a_4 =$ $\frac{1}{2}[(665857/470832)+2(470832/665857)]$ $= (886731088897/627013566048)$
Baudhayana's Method	
$a_1 = 1+\frac{1}{3}+\frac{1}{4}\cdot\frac{1}{3}-\frac{1}{34}\cdot\frac{1}{4}\cdot\frac{1}{3}-\frac{1}{1154}\cdot\frac{1}{34}\cdot\frac{1}{4}\cdot\frac{1}{3}$	
$k_3 = (34)^2-2 = 1154$ $a_3 = 1+\frac{1}{3}+\frac{1}{4}\cdot\frac{1}{3}-\frac{1}{34}\cdot\frac{1}{4}\cdot\frac{1}{3}-\frac{1}{1154}\cdot\frac{1}{34}\cdot\frac{1}{4}\cdot\frac{1}{3}$	
$k_4 = (1154)^2-2 = 1331714$ $a_4 = 1+\frac{1}{3}+\frac{1}{4}\cdot\frac{1}{3}-\frac{1}{34}\cdot\frac{1}{4}\cdot\frac{1}{3}-\frac{1}{1154}\cdot\frac{1}{34}\cdot\frac{1}{4}\cdot\frac{1}{3}-\frac{1}{1331714}\cdot\frac{1}{1154}\cdot\frac{1}{34}\cdot\frac{1}{4}\cdot\frac{1}{3}$	

Figure B.6   Comparison of D&A with Baudhayana's method

However, Baudhayana's method uses significantly fewer computations (in addition, of course, to the drawings either on paper or in one's mind). For example, look at the table in Figure B.6, which compares the methods for the first four approximations. For Baudhayana's method at the n-th stage let k_n denote the number of thin pieces added into the missing square and let c_n denote the correction term that is added.

Notice that the (10-digit) calculator reaches its maximum accuracy at the third stage. At this stage the Baudhayana method obtained more accuracy (it can be checked that it is accurate to 12 digits) and the only computation required was $(34)^2 - 2 = 1154$, which can easily be accomplished by hand. Baudhayana's approximations are numerically identical to those attained in the D&A method using fractions, but again with significantly fewer computations. Of course, Baudhayana's method has this efficiency only if you do not change Baudhayana's representation of the approximation into decimals or into standard fractions. At the fourth stage the Baudhayana method is accurate to less than

$$2[(1331714^2 - 2)(1331714)(1154)(34)(4)(3)]^{-1}$$

or roughly 24-digit accuracy with the only calculation needed being

$$(1154)^2 - 2 = 1331714.$$

Notice that in Baudhayana's fourth representation of the *saviśeṣa* for the square root of 2,

$$\sqrt{2} \approx$$

$$1 + \frac{1}{3} + \frac{1}{4} \cdot \frac{1}{3} - \frac{1}{34} \cdot \frac{1}{4} \cdot \frac{1}{3} - \frac{1}{1154} \cdot \frac{1}{34} \cdot \frac{1}{4} \cdot \frac{1}{3} - \frac{1}{1,331,714} \cdot \frac{1}{1154} \cdot \frac{1}{34} \cdot \frac{1}{4} \cdot \frac{1}{3},$$

the unit is first divided into 3 parts and then each of these parts into 4 parts and then each of these parts into 1154 parts and each of these parts into 133,174 parts. Notice the similarity of this to standard U.S. linear measure where a mile is divided into 8 furlongs and a furlong into 220 yards and a yard into 3 feet and a foot into 12 inches. Other traditional systems of units work similarly except for the metric systems, where the division is always by 10. Also, some carpenters I know, when they have a measurement of $2\frac{7}{16}$ inches, are likely to work with it as $2 + \frac{1}{2} - \frac{1}{8}\frac{1}{2}$, or 2 inches plus a half inch minus an eighth of that half — this is a clearer image to hold onto and work with. It is easier to have an image of the length of $\sqrt{2}$ from Baudhayana's approximation than it is from the D&A's (886731088897/627013566048).

CONCLUSIONS

Baudhayana's method cannot even come close to the D&A method in terms of ease of use with a computer and its applicability to finding the square root of any number. However, the *Sulbasutram* contains many powerful techniques, which, in specific situations, have a power and efficiency that is missing in more general techniques. Numerical computations with the decimal system in either fixed-point or floating-point form has many well-known problems. (See, for example, [**RN:** Turner].) Perhaps we will be able to learn something from the (apparently) first geometry text in the world and devise computational procedures that combine geometry and numerical techniques.

ANNOTATED BIBLIOGRAPHY

> A (cut) Short Story:
> Peter wanted to know the names of the birds.
> He read a book and learned the names of the birds.
> Peter wanted to learn how to swim.
> He read a book and drowned.
> — from E.C. Basar, R.A. Bonic, et al, *Studying Freshman Calculus*, Lexington, MA: D.C. Heath and Co., 1976

This bibliography consists of books (and other items) that are referenced in this book, books used in the writing this book, or books that I have read that I think readers may be interested in. There is a more complete (and frequently updated) bibliography with more than 300 listings on the web at

http://www.math.cornell.edu/~dwh/biblio

The section names below correspond to a subset of the section names in the bibliography on the web.

The annotations in quotes are taken from the introductory matter, cover, or dust jacket of the book being annotated.

AT ANCIENT TEXTS

al'Khowarizmi, *Algebra*

This is the world's first algebra text. An English translation is contained in Karpinski, L.C., ed., *Robert of Chester's Latin Translation of Al'Khowarizmi's Algebra*, New York: Macmillan, 1915.

Apollonius of Perga, *Treatise on Conic Sections*, T.L. Heath, ed., New York: Dover, 1961.

This is the standard work on conic sections from the Greek world.

Apollonius of Perga, *On Cutting Off a Ratio*, E.M. Macierowski, trans., R.H. Schmidt, ed., Fairfield: The Golden Hind Press, 1987.

"An attempt to recover the original argumentation through a critical translation of the two extant medieval Arabic manuscripts."

Baudhayana, *Sulbasutram*, G. Thibaut, trans., S. Prakash & R. M. Sharma, ed., Bombay: Ram Swarup Sharma, 1968.

This is translated from the Sanskrit manual for the construction of altars. The beginning of the book contains a textbook of the geometry needed for the construction of the altars — this beginning section is apparently the oldest surviving geometry textbook. See Appendix B and Chapter 13.

Bonasoni, Paolo, *Algebra Geometrica*, R. H. Schmidt, trans., Annapolis: The Golden Hind Press, 1985.

"being the only known work of this nearly forgotten Renaissance mathematician (excepting a still unpublished treatise on the division of circles)."

Cardano, Girolamo, *The Great Art or the Rules of Algebra*, T.R. Witmer, ed., Cambridge: MIT Press, 1968.

This is the book that first describes algebraic algorithms for solving most cubic equations.

Descartes, Rene, *The Geometry of Rene Descartes*, D. Eugene, M.L. Latham, trans., New York: Dover Publications, Inc., 1954.

This the book in which Descartes develops the use of what we now call *Cartesian coordinates* for the study of curves.

Euclid, *Elements*, T.L. Heath, ed., New York: Dover, 1956.

This is the edition of Euclid's *Elements* to which one is usually referred. Heath has added a large collection of very useful historical and philosophical notes.

Euclid, *Optics*, H. E. Burton, trans., *Journal of the Optical Society of America*, vol. 35, no. 5, pp. 357–72, 1945.

This is a translation of Euclid's work that contains the elements of what we now call projective geometry.

Euclid, *Phaenomena*, in *Euclidis opera omnia*, Heinrich Menge, ed., Lipsiae: B.G. Teubneri, 1883-1916.

A work on astronomy that discusses aspects of spherical geometry.

Guthrie, Kenneth, *The Pythagorean Sourcebook and Library*, Grand Rapids: Phanes Press, 1987.

"An Anthology of Ancient Writings Which Relate to Pythagoras and Pythagorean Philosophy."

Khayyam, Omar, *Algebra*, D.S. Kasir, ed., New York: Columbia Teachers College, 1931 (and New York: A.M.S. Press, 1972).

In this book Khayyam gives geometric techniques for solving cubic equations.

Khayyam, Omar, *Risâla fî sharh mâ ashkala min musâdarât Kitâb 'Uglîdis*, A.I. Sabra, ed., Alexandria, Egypt: Al Maaref, 1961.

This is the Arabic original of Khayyam's discussions of non-Euclidean geometry. Translated in A. R. Amir-Moez, "Discussion of Difficulties in Euclid" by Omar ibn Abrahim al-Khayyami (Omar Khayyam), *Scripta Mathematica*, 24 (1958–59), pp. 275–303.

Khayyam, Omar, a paper (no title).

Translated in A. R. Amir-Moez, "A Paper of Omar Khayyam," *Scripta Mathematica*, 26(1963), pp. 323–37. In this paper Khayyam discusses algebra in relation to geometry.

Plato, *The Collected Dialogues*, Edith Hamilton and Huntington Carns, eds., Princeton, NJ: Bollinger, 1961.

Plato discusses mathematical ideas in many of his dialogues.

Proclus, *Proclus, A Commentary on the First Book of Euclid's Elements*, Glenn R. Morrow, trans., Princeton: Princeton University Press, 1970.

These commentaries by Proclus (Greek, 410–85) are a source of much of our information about the thinking of mathematicians toward the end of the Greek era.

Saccheri, Girolamo, *Euclides Vindicatus*, G.B. Halsted, ed. and trans., New York: Chelsea Pub. Co., 1986.
"In this book Girolamo Saccheri set forth in 1733, for the first time ever, what amounts to the axiom systems of non-Euclidean geometry." It is not mentioned in this volume that Saccheri borrowed many ideas from Khayyam's *Risâla fî sharh mâ ashkala min musâdarât Kitâb 'Uglîdis*.

Theon of Smyrna, *Mathematics Useful for Understanding Plato*, R. and D. Lawlor, trans., San Diego: Wizards Bookshelf, 1978.
"This work appears to have been a text book intended for students who were beginning a study of the works of Plato. In its original form there were five sections: 1) Arithmetic 2) Plane Geometry 3) Stereometry (solid geometry) 4) Music 5) Astronomy. Sections 2 and 3 on Geometry have been lost while the others remain in their entirety and are presented here." The section on Astronomy contains discussions of the shape of space.

Thomas, Ivor, trans., *Selections Illustrating the History of Greek Mathematics* , Cambridge, MA: Harvard University Press, 1951.
A collection of primary sources.

CG Computers and Geometry

The Geometer's Sketchpad: Dynamic Geometry for the 21st Century, Key Curriculum Press.
A program running on Windows or Mac platforms that allows you to construct geometric drawing with points, lines, and circles and then dynamically vary constituent parts.

Richter-Gerbert, Jürgen, and Ulrich H. Kortenkamp, *Cinderella*: *The Interactive Geometry Software*, Heidelberg: Springer-Verlag, 1999.
A *Java*-based dynamic geometry software.

Taylor, Walter F., *The Geometry of Computer Graphics*, Pacific Grove, CA: Wadsworth & Brooks/Cole Advanced Books & Software, 1992.

> "This book is a direct presentation of elementary analytic and projective geometry, as modeled by vectors and matrices and as applied to computer graphics."

DG DIFFERENTIAL GEOMETRY

Bloch, Ethan D., *A First Course in Geometric Topology and Differential Geometry*, Boston: Birkhauser, 1997.

> This book contains the topological classification and differential geometry of surfaces.

Casey, James, *Exploring Curvature*, Wiesbaden: Vieweg, 1996.

> A truly delightful book full of "experiments" to physically explore curvature of curves and surfaces.

do Carmo, Manfredo, *Differential Geometry of Curves and Surfaces*, Englewood Cliffs, NJ: Prentice Hall, 1976.

> An undergraduate level text.

Dodson, C.T.J., and T. Poston, *Tensor Geometry*, London: Pitman, 1979.

> A very readable but technical text using linear (affine) algebra to study the local intrinsic geometry of spaces leading up to and including the geometry of the theory of relativity.

Gauss, C.F., *General Investigations of Curved Surfaces*, Hewlett, NY: Raven Press, 1965.

> A translation into English of Gauss' early papers on surfaces.

Gray, A., *Modern Differential Geometry of Curves and Surfaces*, Akron: CRC, 1993.

> This is a very extensive book based on computations using Mathematica©.

Henderson, David W., *Differential Geometry: A Geometric Introduction*, Upper Saddle River, NJ: Prentice Hall, 1998.

"In this book we will study a foundation for differential geometry based not on analytic formalisms but rather on these underlying geometric intuitions."

Koenderink, Jan J., *Solid Shape*, Cambridge: M.I.T. Press, 1990.

Written for engineers and applied mathematicians, this is a discussion of the extrinsic properties of three-dimensional shapes. There are connections with applications and a nice section called "Your way into the literature."

Kreyszig, Erwin, *Mathematical Expositions No. 11: Differential Geometry*, Toronto: University of Toronto Press, 1959.

"This book provides an introduction to the differential geometry of curves and surfaces in three-dimensional Euclidean space... In the theory of surfaces we make full use of the tensor calculus, which is developed as needed."

McCleary, John, *Geometry from a Differential Viewpoint*, Cambridge, UK: Cambridge University Press, 1994.

"The text serves as both an introduction to the classical differential geometry of curves and surfaces and as a history of ... the hyperbolic plane."

Morgan, Frank, *Riemannian Geometry: A Beginner's Guide*, Boston: Jones and Bartlett, 1993.

An accessible guide to Riemannian geometry including a chapter on the theory of relativity and the calculation of the precession in the orbit of Mercury.

Penrose, Roger, "The Geometry of the Universe," *Mathematics Today*, Lynn Steen, ed., New York: Springer-Verlag, 1978.

An expository discussion of the geometry of the universe.

Rovenski, V.Y., *Geometry of Curves and Surfaces with MAPLE*, Boston: Birkhäuser, 1998.

"This concise text on geometry with computer modeling presents some elementary methods for analytical modeling and visualization on curves and surfaces."

Santander, M., "The Chinese South-Seeking chariot: A simple mechanical device for visualizing curvature and parallel transport," *American Journal of Physics*, vol. 60, no. 9, pp. 782–87, 1992.

"An old mechanical device, the Chinese South-Seeking chariot, presumably designed to work on a flat plane, is shown to perform parallel transport on arbitrary surfaces. Its use affords experimental demonstration and even numerical checking (within a reasonable accuracy) of all the features of curvature and parallel transport of vectors in a two-dimensional surface."

Spivak, Michael, *A Comprehensive Introduction to Differential Geometry*, Wilmington, DE: Publish or Perish, 1979.

In five (!) volumes Spivak relates the subject back to the original sources. Volume V contains an extensive bibliography (to 1979).

Stahl, Saul, *The Poincaré Half-Plane*, Boston: Jones and Bartlett Publishers, 1993.

This text is an analytic introduction to some of the ideas of intrinsic differential geometry starting from the Calculus.

Thurston, William, *Three-Dimensional Geometry and Topology*, Vol. 1, Princeton, NJ: Princeton University Press, 1997.

This is a detailed excursion through the geometry and topology of 2- and 3-manifolds. "The style of exposition in this book is intended to encourage the reader to pause, to look around and to explore."

Weeks, Jeffrey, *The Shape of Space*, New York: Marcel Dekker, 1985.

An elementary but deep discussion of the geometry on different two- and three-dimensional spaces.

DI   DISSECTIONS

Boltyanski (Boltianskii), Vladimir G., *Hilbert's Third Problem*, New York: John Wiley & Sons, 1978.

A discussion of dissections on the plane, sphere, and hyperbolic spaces.

Eves, Howard, *A Survey of Geometry,* Vol. 1, Boston: Allyn & Bacon, 1963.

A textbook that contains an extensive coverage of the dissection theory of polygons.

Frederickson, Greg, *Dissections*: *Plane and Fancy,* New York: Cambridge University Press, 1997.

This book is a collection of interesting dissection puzzles, old and new, and is an instructive manual on the art and science of geometric dissections.

Ho, Chung-Wu, "Decomposition of a Polygon into Triangles," *Mathematical Gazette,* vol. 60, pp.132–34, 1976.

This article contains a proof that all planar polygons can be dissected into triangles and discusses the many mistakes made by other (many well-known) authors in their "proofs" of the same result.

Sah, C.H., *Hilbert's Third Problem: Scissors Congruence,* London: Pitman, 1979.

A detailed discussion of the three-dimensional dissections.

DS DIMENSIONS AND SCALE

Abbott, Edwin A., *Flatland,* New York: Dover Publications, Inc., 1952.

A fantasy about two-dimensional beings in a plane encountering the third dimension.

Banchoff, Thomas, and John Wermer, *Beyond the Third Dimension: Geometry, Computer Graphics, and Higher Dimensions*, New York: Springer-Verlag, 1983.

"This book treats a number of themes that center on the notion of dimensions, tracing the different ways in which mathematicians and others have met them in their work."

Burger, Dionys, *Sphereland,* New York: Thomas Y. Crowell Co., 1965.

A sequel to Abbott's *Flatland.*

Kohl, Judith, and Herbert Kohl, *The View from the Oak: The Private Worlds of Other Creatures*, New York: Sierra Club Books/Charles Scribner's Sons, 1977.

> This delightful book describes the various experiential worlds of different creatures and is a good illustration of intrinsic ways of thinking. Included are differing dimensions and scales of these worlds.

Morrison, Phillip, and Phylis Morrison, *Powers of Ten: About the Relative Size of Things in the Universe*, New York: Scientific American Books, Inc., 1982.

> A beautiful book (and a video with the same title) that starts with a square meter on earth and then zooms out and in by powers of ten describing and illustrating at each power of ten what can be seen until it reaches (by zooming out) vast stretches of empty space in the universe or (by zooming in) the empty space within elementary particles.

Rucker, Rudy, *The Fourth Dimension*, Boston: Houghton Mifflin Co., 1984.

> A history and description of various ways that people have considered the fourth dimension.

Rucker, Rudy, *Geometry, Relativity and the Fourth Dimension*, New York: Dover, 1977.

> "[The author's] goal has been to present an intuitive picture of the curved space-time we call home."

GC GEOMETRY IN DIFFERENT CULTURES

Albarn, K., Jenny Miall Smith, Stanford Steele, Dinah Walker, *The Language of Pattern*, New York: Harper & Row, 1974.

> An inquiry inspired by Islamic decoration.

Ascher, Marcia, *Ethnomathematics: A Multicultural View of Mathematical Ideas*, Pacific Grove, CA: Brooks/Cole, 1991.

> A mostly anthropological look at the mathematics indigenous to several ancient cultures.

Bain, George, *Celtic Arts: The Methods of Construction*, London: Constable, 1977.

A description of the construction of Celtic patterns and designs.

Data, *The Science of the Sulba*, Calcutta: University of Calcutta, 1932.

A discussion of the mathematics in the *Sulbasutram* and traditional Hindi society.

Gerdes, Paulus, *Women, Art and Geometry in Southern Africa*, Trenton: Africa World Press, Inc., 1998.

"The main objective of the book *Women, Art and Geometry in Southern Africa* is to call attention to some mathematical aspects and ideas incorporated in the patterns invented by women in Southern Africa."

Gerdes, Paulus, *Geometry From Africa*: *Mathematical and Educational Explorations*, Washington: Mathematical Association of America, 1999.

"... we learn of the diversity, richness, and pleasure of mathematical ideas found in Sub-Saharan Africa. From a careful reading and working through this delightful book, one will find a fresh approach to mathematical inquiry as well as encounter a subtle challenge to Eurocentric discourses concerning the when, where, who, and why of mathematics."

Pinxten, R., Ingrid van Dooren, Frank Harvey, *The Anthropology of Space*, Philadelphia: University of Pennsylvania Press, 1983.

Concepts of geometry and space in the Navajo culture.

Zaslavsky, Claudia, *Africa Counts*, Boston: Prindle, Weber, and Schmidt, Inc., 1973.

A presentation of the mathematics in African cultures.

HI HISTORY

Berggren, *Episodes in the Mathematics of Medieval Islam*, New York: Springer-Verlag, 1986.

Describes many examples that are difficult to find elsewhere of the mathematical contributions from medieval Islam.

Calinger, Ronald, *Classics of Mathematics*, Englewood Cliffs, NJ: Prentice Hall, 1995.

Mostly a collection of original sources in Western mathematics.

Carroll, Lewis, *Euclid and His Modern Rivals*, New York: Dover Publications, Inc., 1973.

Yes! Lewis Carroll of *Alice in Wonderland* fame was a geometer. This book is written as a drama; Carroll has Euclid defending himself against modern critics.

Critchlow, K., *Time Stands Still*, London: Gordon Fraser, 1979.

This book describes evidence of prehistoric scientific and mathematical knowledge.

Eves, Howard, *Great Moments in Mathematics (after 1650)*, Dolciani Mathematical Expositions, vol. 7, Washington, DC: M.A.A., 1981.

This small book contains 20 lectures: 2 on non-Euclidean geometry and one on Klien's "Erlanger Program" which set out to delineate various aspects of mathematics (including especially geometry) according to which transformations left invarient important properties.

Fauvel, John, and Jeremy Gray, *The History of Mathematics: A Reader*, London: Macmillan Press, 1987.

"The selection of readings has been made for students of the Open University course MA290 *Topics in the History of Mathematics* ..."

Gray, Jeremy, *Ideas of Space: Euclidean, Non-Euclidean and Relativistic*, 2nd edition, Oxford: Oxford University Press, 1989.

A mostly historical account of Euclidean, non-Euclidean, and relativistic geometry.

Heath, T.L., *Mathematics in Aristotle,* Oxford: Clarendon Press, 1949.

Discusses the mathematical contributions of Aristotle.

Heath, T.L., ed., *Euclid: The Thirteen Books of the Elements*, New York: Dover, 1956.

This is the edition of Eulid's *Elements* to which one is usually referred. Heath has added a large collection of very useful historical and philosophical notes. His notes are more extensive than Euclid's text.

Helbron, J.L., *Geometry Civilized: History, Culture, and Technique*, Oxford: Clarendon Press, 2000.
> "For many centuries, geometry was part of high culture as well as an instrument of practical utility."

Joseph, George, *The Crest of the Peacock*, New York: I.B. Tauris, 1991.
> A non-Eurocentric view of the history of mathematics.

Katz, Victor J., *A History of Mathematics: An Introduction*, Reading, MA: Addison-Wesley Longman, 1998.
> "... designed for junior or senior mathematics majors who intend to teach in college or high school and thus concentrates on the history of those topics typically covered in an undergraduate curriculum or in elementary or high school."

Kline, Morris, *Mathematical Thought from Ancient to Modern Times*, Oxford: Oxford University Press, 1972.
> A complete Eurocentric history of mathematical ideas including differential geometry (mostly the analytic side).

Laubenbacher, Reinhard, and David Pengelley, *Mathematical Expeditions: Chronicles by the Explorers*, New York: Springer, 1999.
> Contains a 53-page chapter on "Geometry: The Parallel Postulate."

Newell, Virginia K., ed., *Black Mathematicians and Their Works*, Ardmore, PA: Dorrance, 1980.
> A discussion of (mostly American) black mathematicians and their mathematics.

Richards, Joan, *Mathematical Visions*, Boston: Academic Press, 1988.
> "The pursuit of geometry in Victorian England."

Rosenfeld, B.A., *A History of Non-Euclidean Geometry*, New York: Springer-Verlag, 1989.
> An extensive history of non-Euclidean geometry based on original sources.

Seidenberg, A., "The Ritual Origin of Geometry," *Archive for the History of the Exact Sciences*, vol. 1, pp. 488–527, 1961.
In this article Seidenberg makes the case that much geometry originated from the needs of various religious rituals.

Snyder, John P., *Flattening the Earth: Two Thousand Years of Map Projections*, Chicago: University of Chicago Press, 1993.
A history and mathematical description of numerous map projections of the sphere.

Toth, I., "Non-Euclidean Geometry before Euclid," *Scientific American*, vol. 251, 1969.
Discusses the evidence of non-Euclidean geometry before Euclid.

Valens, Evans G., *The Number of Things: Pythagoras, Geometry and Humming Strings*, New York: E.P. Dutton and Company, 1964.
This is a book about ideas and is not a textbook. Valens leads the reader through dissections, golden mean, relations between geometry and music, conic sections, etc.

van der Waerden, B.L., *Science Awakening I: Egyptian, Babylonian, and Greek Mathematics*, Princeton Junction, NJ: The Scholar's Bookshelf, 1975.
"It is the intention to make this book scientific, but at the same time accessible to any one who has learned some mathematics in school and in college, and who is interested in the history of mathematics."

MP MODELS, POLYHEDRA

Barr, Stephen, *Experiments in Topology*, New York: Crowell, 1964.
Experimental topology that goes beyond the Möbius Band.

Cundy, M.H., and A.P. Rollett, *Mathematical Models*, Oxford: Clarendon, 1961.
Directions on how to make and understand various geometric models.

Ehrenfeucht, Aniela, *The Cube Made Interesting*, Waclaw Zawadowski, trans., New York: Pergamon Press, 1964.

"This book arose from popular scientific talks to teachers and school children." The discussion is illustrated with 3-D pictures using special glasses.

Lénárt, István, *Lénárt Sphere Construction Materials: Construction Materials for Another World of Geometry*, Berkeley: Key Curriculum Press, 1996.

This is a kit consisting of a transparent sphere, a spherical compass, and a spherical "straight edge" that doubles as a protractor. Other accessory materials are available.

Lyusternik, L.A., *Convex Figures and Polyhedra*, Boston: Heath, 1966.

A detailed but elementary study of convex figures.

Row, T. Sundra, *Geometric Exercises in Paper Folding*, New York: Dover, 1966.

How to produce various geometric constructions merely by folding a sheet of paper.

Senechal, Marjorie, and George Fleck, eds., *Shaping Space: A Polyhedral Approach*, Design Science Collection, Boston: Birkhauser, 1988.

This book is an accessible "exploration of the world of polyhedra, beginning with [an introduction] and concluding with an examination of the significance of polyhedral models in contemporary science and a survey of some recent advances and unsolved problems in mathematics."

Na Nature

Cook, T.A., *The Curves of Life*, New York: Dover Publications, 1979.

Subtitle: *Being an Account of Spiral Formations and their Applications to Growth in Nature, to Science, and to Art.*

Hildebrandt, Stefan, and Anthony Tromba, *Mathematics and Optimal Form*, New York: Scientific American Books, Inc., 1985.

"Combining striking photographs with a compelling text, authors ... give us a thoughtful account of the symmetry and regularity of nature's forms and patterns."

Kohl, Judith, and Herbert Kohl, *The View from the Oak: The Private Worlds of Other Creatures*, New York: Sierra Club Books/Charles Scribner's Sons, 1977.

This delightful books describes the various experiential worlds of different creatures and is a good illustration of intrinsic ways of thinking.

Mandelbrot, Benoit B., *The Fractal Geometry of Nature*, New York: W.H. Freeman and Company, 1983.

The book that started the popularity of fractal geometry.

McMahon, Thomas, and James Bonner, *On Size and Life*, New York: Scientific American Library, 1983.

A geometric discussion of the shapes and sizes of living things.

Thompson, D'Arcy, *On Growth and Form*, Cambridge: Cambridge University Press, 1961.

A classic on the geometry of the natural world.

NE NON-EUCLIDEAN GEOMETRIES (MOSTLY HYPERBOLIC)

Bonola, Roberto, *Non-Euclidean Geometry: A Critical and Historic Study of its Developments, and "The Theory of Parallels" by Nicholas Lobachevski with a supplement containing "The Science of Absolute Space" by John Bolyai*, New York: Dover, 1955.

"Bonola's Non-Euclidean Geometry is an elementary historical and critical study of the development of that subject."

Greenberg, Marvin J., *Euclidean and Non-Euclidean Geometries: Development and History*, New York: Freeman, 1980.

This is a very readable textbook that includes some philosophical discussions.

Hilbert, David, "Über Flächen von konstanter gausscher Krümmung," *Transactions of the A.M.S.*, pp. 87–99, 1901.

Hilbert proves here that the hyperbolic plane does not have a real analytic (or C^4) isometric embedding onto a closed subset of Euclidean 3-space.

Kuiper, Nicolas, "On C^1-isometric embeddings, ii," *Nederl. Akad. Wetensch. Proc. Ser. A*, pp. 683–89, 1955.

Kuiper shows that there is a C^1 isometric embedding of the hyperbolic plane onto a closed subset of Euclidean 3-space.

Millman, Richard S., and George D. Parker, *Geometry: A Metric Approach with Models*, New York: Springer-Verlag, 1981.

A modern formal axiomatic approach.

Milnor, Tilla, Efimov's theorem about complete immersed surfaces of negative curvature, *Advances in Math.*, vol. 8, pp. 474–543, 1972.

Milnor proves that it is impossible to have a C^2 isometric embedding of the hyperbolic plane onto a closed subset of Euclidean 3-space.

Petit, Jean-Pierre, *Euclid Rules OK? The Adventures of Archibald Higgins*, London: John Murray, 1982.

A pictorial, visual tour of non-Euclidean geometries.

Schwerdtfeger, Hans, *Geometry of Complex Numbers: Circle Geometry, Moebius Transformation, Non-Euclidean Geometry*, New York: Dover Publications, Inc., 1979.

This book uses complex numbers to analyze inversions in circles and then their relationship to hyperbolic geometry.

Singer, David A., *Geometry: Plane and Fancy*, New York: Springer, 1998.

"This book is about ... the idea of curvature and how it affects the assumptions about and principles of geometry."

Stahl, Saul, *The Poincaré Half-Plane*, Boston: Jones and Bartlett Publishers, 1993.

This text is an analytic introduction to some of the ideas of intrinsic differential geometry starting from the Calculus.

Trudeau, Richard J., *The Non-Euclidean Revolution*, Boston: Birkhäuser, 1987.

"Trudeau's book provides the reader with a non-technical description of the progress of thought from Plato and Euclid to Kant, Lobachevsky, and Hilbert."

Zage, "The Geometry of Binocular Visual Space," *Mathematics Magazine*, vol. 53, no. 5, pp. 289–94, 1980.

"... we relate the results of experiments in binocular vision to geometric models to arrive at the conclusion that the geometry of binocular visual space is [...] hyperbolic."

PH PHILOSOPHY

Benacerraf, Paul, and Hilary Putman, *Philosophy of Mathematics: Selected Readings*, Cambridge: Cambridge University Press, 1964.

An interesting and useful selection of readings.

Lakatos, I., *Proofs and Refutations*, Cambridge: Cambridge University Press, 1976.

A deep but accessible book that uses an imaginary classroom dialogue in which the actual historical words of mathematicians are used to explore the evolving nature of mathematical ideas and to support the author's *quasi empirical* view of mathematics.

Rucker, Rudy, *Infinity and the Mind: The Science and Philosophy of the Infinite*, Boston: Birkhauser, 1982.

"This book discusses every kind of infinity: potential and actual, mathematical and physical, theological and mundane. Talking about infinity leads to many fascinating paradoxes. By closely examining these paradoxes we learn a great deal about the human mind, its powers, and its limitations."

Tymoczko, Thomas, *New Directions in the Philosophy of Mathematics*, Boston: Birkhauser, 1986.

An updated (to 1986) collection of readings.

RN REAL NUMBERS

Epstein, Richard L., and Walter A. Carnielli, *Computability: Computable Functions, Logic, and the Foundations of Mathematics*, Pacific Grove, CA: Wadsworth & Brooks/Cole, 1989.

"This book... deals with a very basic problem: What is computable?"

Simpson, "The Infidel Is Innocent," *The Mathematical Intelligencer*, vol. 12, pp.42–51, 1990.

An accessible exposition of the nonstandard reals.

Turner, Peter R., "Will the 'Real' Real Arithmetic Please Stand Up?", *Notices of the A.M.S.*, vol. 38, pp.298–304, 1991.

An article about various finite representations of real numbers used in computing.

SE SURVEYS AND GENERAL EXPOSITIONS

Coxeter, H.S.M., and S.L. Greitzer, *Geometry Revisited*, New Mathematics Library 19, New York: The L.W. Singer Company, 1967.

"Using whatever means will best suit our purposes, let us revisit Euclid. Let us discover for ourselves a few of the newer results. Perhaps we may be able to recapture some of the wonder and awe that our first contact with geometry aroused."

Davis, P.J., and R. Hersh, *The Mathematical Experience*, Boston: Birkhauser, 1981.

> A very readable collection of essays by two present-day mathematicians. I think every mathematics major should own this book.

Gorini, Catherine, ed., *Geometry at Work*, MAA Notes, vol. 53, Washington, DC: MAA, 2000.

> A varied collection of writings on applications of geometry.

Hilbert, David, and S. Cohn-Vossen, *Geometry and the Imagination*, New York: Chelsea Publishing Co., 1983.

> They state "it is our purpose to give a presentation of geometry, as it stands today [1932], in its visual, intuitive aspects." It includes an introduction to differential geometry, symmetry, and patterns (they call it "crystallographic groups"), and the geometry of spheres and other surfaces. Hilbert is the most famous mathematician of the first part of this century.

SG SYMMETRY AND GROUPS

Budden, F.J., *Fascination of Groups*, Cambridge: Cambridge University Press, 1972.

> This is a fascinating book that relates algebra (groups) to geometry, music, and so forth, and has a nice description of symmetry and patterns.

Grünbaum, Branko, and G.C. Shepard, *Tilings and Patterns*, New York: W.H. Freeman, 1987.

> A 700-page book detailing what is known about plane tilings and patterns.

Lyndon, Roger C., *Groups and Geometry*, New York: Cambridge University Press, 1985.

> "This book is intended as an introduction, demanding a minimum of background, to some of the central ideas in the theory of groups and in geometry. It grew out of a course for advanced undergraduates and beginning graduate students."

Macgillavry, Caorline H., *Symmetry Aspects of M.C. Escher's Periodic Drawings*, Utrecht: Published for the International Union of Crystallography by A. Oosthoek's Uitgeversmaatschappij NV, 1965.

> This volume describes a scheme for classifying periodic patterns with colors, using as examples Escher's drawings.

Martin, George E., *Transformation Geometry: An Introduction to Symmetry*, New York: Springer-Verlag, 1982.

> "Our study of the automorphisms of the plane and of space is based on only the most elementary high-school geometry. In particular, group theory is *not* a prerequisite here. On the contrary, this modern approach to Euclidean geometry gives the concrete examples that are necessary to appreciate an introduction to group theory."

Montesinos, José María, *Classical Tessellations and Three-Manifolds*, New York: Springer Verlag, 1985.

> "This book explores a relationship between classical tessellations and three-manifolds."

Robertson, Stewart A., *Polytopes and Symmetry*, New York: Cambridge University Press, 1985.

> "These notes are intended to give a fairly systematic exposition of an approach to the symmetry classification of convex polytopes that casts some fresh light on classical ideas and generates a number of new theorems."

Weyl, Hermann, *Symmetry*, Princeton, NJ: Princeton University Press, 1952.

> A readable discussion of all mathematical aspects of symmetry, especially its relation to art and nature — nice pictures. Weyl is a leading mathematician of this century.

Yale, Paul B., *Geometry and Symmetry*, New York: Dover, 1988.

> "This book is an introduction to the geometry of Euclidean, affine and projective spaces with special emphasis on the important groups of symmetries of these spaces."

SP SPHERICAL AND PROJECTIVE GEOMETRY

Albert, A. Adrian, and Reuben Sandler, *An Introduction to Finite Projective Planes*, New York: Holt, Rinehart and Winston, 1968.

"In this book the authors have endeavored to introduce the subject of finite projective planes as it has developed during the last twenty years."

Coxeter, H.S.M., *The Real Projective Plane*, New York: Cambridge University Press, 1955.

"This introduction to projective geometry can be understood by anyone familiar with high-school geometry and algebra. The restriction to real geometry of two dimensions makes it possible for every theorem to be illustrated by a diagram."

Lénárt, István, *Non-Euclidean Adventures on the Lénárt Sphere: Activities comparing planar and spherical geometry*, Berkeley: Key Curriculum Press, 1996.

This is a manual for high school teachers using the Lénárt Sphere.

Todhunter, Isaac, *Spherical Trigonometry*, London: Macmillan, 1886.

All you want to know, and more, about trigonometry on the sphere. Well-written with nice discussions of surveying.

TG TEACHING GEOMETRY

Malkevitch, Joseph, *Geometry's Future second edition*, USA: COMAP, 1991.

Proceedings of a COMAP conference "of a small group of geometers to study what could be done to revitalize geometry in our colleges, and what effects this might have on the teaching of geometry in general."

Mammana, Carnelo, and Vinicio Villani, ed., *Perspectives on the Teaching of Geometry for the 21st Century: ICMI Study*, Dortrecht, Netherlands: Kluwer, 1998.

Contains papers and the edited summaries and conclusions from the ICMI Study.

Traylor, Reginald, *Creative Teaching: Heritage of R.L. Moore*, Houston: University of Houston, 1972. Also at http://at.yorku.ca/i/a/a/b/21.dir/

> A mathematical teaching biography of R.L. Moore and list of his mathematical descendents.

Zimmermann, Walter and Cunningham, Steve, *Visualization in Teaching and Learning Mathematics*, USA: Mathematical Association of America, 1991.

> "A project sponsored by the Committee on Computers in Mathematics Education of the M.A.A."

Tp Topology

Arnold, B.H., *Intuitive Concepts in Elementary Topology*, Englewood Cliffs: Prentice-Hall, Inc., 1962.

> "Topology is presented here from the intuitive, rather than the axiomatic viewpoint. Some concepts are introduced, discussed and used informally, on the basis of the student's experience."

Blackett, Donald W., *Elementary Topology*, New York: Academic Press, 1967.

> Contains a combinatorial-based proof of the classification of 2-manifolds.

Cairns, Stewart Scott, *Introductory Topology*, New York: The Ronald Press Company, 1961.

> "This book is the culmination of repeatedly revised sets of class notes used by the author in teaching introductory topology courses. Its purpose is to progress as far as practicable into the fundamental concepts and the principal results of homology theory, both in their combinatorial development and in their application to topological spaces."

Francis, G.K., *A Topological Picturebook*, New York: Springer Verlag, 1987.

> Francis presents elaborate and illustrative drawings of surfaces and provides guidelines for those who wish to produce such drawings.

Francis, G.K., and Jeffrey R. Weeks, "Conway's ZIP Proof", *American Mathematical Monthly*, vol. 106, May 1999, pages 393–99, May 1999.

A new proof of the classification of (triangulated) surfaces (2-manifolds).

Tx Geometry Texts

Berger, M., *Geometry I & II*, Sole, M. and Levy, trans., New York: Springer-Verlag, 1987.

"This two-volume textbook is the translation of the French book 'Géométrie' originally published in five volumes. It gives a detailed treatment of classical geometry and provides a comprehensive and unified reference source for all the subfields of geometry, including crystallographic groups; affine, Euclidean, Spherical and hyperbolic geometries; projective geometry; geometry of triangles, tetrahedra, circles and spheres; convex sets, convex and regular polyhedra, etc."

Bruni, James V., *Experiencing Geometry*, Wadsworth Publishing Company, Inc., Belmont, CA, 1977.

"This book is meant to be an informal, intuitive *introduction* to geometry. If you glance through the book, you will find none of the formal theorems of proofs you might expect in a mathematics textbook. You will find numerous concrete models and illustrations intended to be a springboard for the discovery of geometric ideas. Through guided observation, experimentation, and the use of your intuition, you will investigate some fundamental geometric concepts. These experiences can serve as preparation for a more formal, abstract, and logically precise study of different kinds of geometry."

Coxeter, H.S.M., *Introduction to Geometry*, New York: Wiley, 1969.

This is a collection of diverse topics including non-Euclidean geometry, symmetry, patterns, and much, much more. Coxeter is one of the foremost living geometers.

Hansen, Vagn Lundsgaard, *Shadows of the circle: Conic Sections, Optimal Figures and Non-Euclidean Geometry*, Singapore: World Scientific, 1998.

"It is my hope that these topics will be an inspiration in connection with teaching of geometry at various levels including upper secondary school and college education."

Jacobs, Harold R., *Geometry*, San Fransisco: W.H. Freeman and Co., 1974.

A high-school-level text based on guided discovery.

Martin, George E., *Geometric Constructions*, New York: Springer, 1998.

A geometry textbook based on ruler and compass constructions.

Pedoe, Dan, *Geometry: A Comprehensive Course*, New York: Dover Publications, Inc., 1970.

"The main purpose of the course was to increase geometrical, and therefore mathematical understanding, and to help students to enjoy geometry. This is also the purpose of my book."

Serra, Michael, *Discovering Geometry: An Inductive Approach*, Berkeley, CA: Key Curriculum Press, 1989.

A high school text that promotes discovery.

Shurman, Jerry, *Geometry of the Quintic*, New York: John Wiley & Sons, Inc., 1997.

An advanced undergraduate text that uses the icosahedron to solve quintic equations. Along the way, he explores the Riemann sphere, group representations, and invariant functions.

Sibley, Thomas Q., *The Geometric Viewpoint: A Survey of Geometries*, Reading, MA: Addison Wesley, 1998.

"Geometry combines visual delights and powerful abstractions, concrete intuitions and general theories, historical perspective and contemporary applications, and surprising insights and satisfying certainty. In this textbook, I try to weave together these facets of geometry. I also want to convey the multiple connections that different topics in geometry have with each other and that geometry has with other areas of mathematics."

Singer, David A., *Geometry: Plane and Fancy*, New York: Springer, 1998.

> "This book is about... the idea of curvature and how it affects the assumptions about and principles of geometry."

Wallace, Edward C. and Stephen F. West, *Roads to Geometry*, Upper Saddle River, NJ: Prentice Hall, Inc., 1998.

> "The goal of this book is to provide a geometric experience which clarifies, extends, and unifies concepts which are generally discussed in traditional high school geometry courses and to present additional topics which assist in gaining a better understanding of elementary geometry."

Un The Physical Universe

Ferris, Timothy, *The Whole Shebang*: *A State-of-the-Universe(s) Report*, New York: Simon & Schuster, 1997.

> "This book aims to summarize the picture of the universe that science has adduced..., and to forecast an exciting if unsettling new picture that may emerge in the near future."

Guth, Alan H., *The Inflationary Universe*: *The Quest for a New Theory of Cosmic Origins*, Reading, MA: Perseus Books, 1997.

> "The inflationary universe is a theory of the 'bang' of the big bang."

Osserman, Robert, *Poetry of the Universe*: *A Mathematical Exploration of the Cosmos*, New York: Anchor Books, 1995.

> "What is the shape of the universe, and what do we mean by the curvature of space? One aim of this book is to make absolutely clear and understandable both the meanings of those questions and the answers to them. Little or no mathematical background is needed..."

Rees, Martin, *Before the Beginning*: *Our Universe and Others*, Reading, MA: Perseus Books, 1997.

"This book presents an individual view on cosmology — how we perceive our universe, what the current debates are about, and the scope and limits of our future knowledge."

Z MISCELLANEOUS

Davis, Phillip, *The Thread*: *A Mathematical Yarn*, Boston: Birkhäuser, 1983.
> Contains a story about the discovery of the mechanism that turns circular motion into straight line motion and thus can be used for drawing a straight line.

Kempe, A.B., *How to Draw a Straight Line*, London: Macmillan, 1877.
> This small book contains a discussion and description of numerous curve-drawing devices including ones that will draw straight lines.

Snyder, John P., *Flattening the Earth*: *Two Thousand Years of Map Projections*, Chicago: University of Chicago Press, 1993.
> A history and mathematical description of numerous map projections of the sphere.

Sobel, Dava, *Longitude*: *The True Story of a Lone Genius Who Solved the Greatest Scientific Problem of His Time*, New York: Penguin Books, 1995.
> An account of the struggles to develop a method for determining the longitude of ships at sea.

INDEX

The numbers in **bold** refer to pages on which there are definitions or statements of the indexed item.

F

flat
  Klein bottle, 212–8
  torus, 212–8
  two-manifold, **208**
formal mathematics, 27
four-space, 265–78
fractals, 9

G

Gauss, 45, 92, 131, 291
Gauss–Bonnet Formula, 91–2, 95–7, 228
geodesic, **17**, 19, 23–4, 33–4, 38–40, 64, 87, 188, 259
  asymptotic, **57**, 106, 135–7
  in $\mathbf{H}^3$, 274
  in $\mathbf{S}^3$, 270
  locally unique, 63
  radial, **57**, 136–7, 194, 198, 201
geodesic rectangular coordinates, **194**, 196
geometric 3-manifold, **288**, 290
geometric two-manifold, **207**
  flat (=Euclidean), **208**
  hyperbolic, **224**
  spherical, **219**
glide reflection, **135**, 142, 145
great 2-sphere, **270**, 273–9, 284
great circle, **16–7**, 18–21, 39, 87, 188, 191, 256–61, 271, 274–8, 281
great hemisphere, **273**
great semicircle, **273**

H

half turn manifold, 295
half-turn, 7, 142
helix, 7, 39
hexagonal 3-torus, 295
Hilbert's Third Problem, 175

holonomy, 79, 85, **88–92**, 95–7, 120, 129, 157–8, 228
homeomorphic, 214
homogeneous, 48
horolation, **135–6**, 145
hyperbola, 240–1, 245
hyperbolic 2-manifold, 224–6
hyperbolic 3-space, 265, **272**
hyperbolic geodesics, 204
hyperbolic geometry, 46, 128
hyperbolic plane, 33, **45–61**, 65, 75, 81–97, 100, 103–6, 110–6, 117–30, 134–6, 144–8, 152, 155–8, 177, 193–205, 207, 219, 225–30, 251, 253, 256, 261, 272, 276, 278, 384, 291–2, 326, 336
  annular, **48–9**, 50–4, 194–5, 198
  crocheted, 49
  geodesics on, 57, 60–1, 194–201
  Poincaré disk model, 203–4
  polyhedral, 51–5
  projective disk model, 205
  radius of, 55–6
  upper half plane model, 194–8
hyperbolic reflections, 200, 204

I

icosahedron, 286
indistinguishable, 12
intrinsic, 17, 21, 37, 41, 60, 87, 230–1, 252, 265
intrinsic (= covariant) derivative, 24, 87
inversion, **180–3**, 199
inversive pair, **180**
isometric
  locally, **41**, 60
isometry (see also symmetry), **3**, **133**, 137, 142–3, 198
isomorphic groups, 149
isomorphic patterns, **138**